Nonlinear Filtering

Kumar Pakki Bharani Chandra
Da-Wei Gu

Nonlinear Filtering

Methods and Applications

Springer

Kumar Pakki Bharani Chandra
GMR Institute of Technology
Rajam, Andhra Pradesh, India

Da-Wei Gu
Department of Engineering
University of Leicester
Leicester, UK

Additional material to this book can be downloaded from http://extras.springer.com.

ISBN 978-3-030-01796-5 ISBN 978-3-030-01797-2 (eBook)
https://doi.org/10.1007/978-3-030-01797-2

Library of Congress Control Number: 2018958347

This Springer imprint is published by the registered company Springer Nature Switzerland AG
The registered company address is: Gewerbestrasse 11, 6330 Cham, Switzerland

To our families

Preface

This book discusses state estimation of nonlinear systems. As commonly recognised, control systems engineering and control systems theory can be traced back to James Watt's development of the flyball governor in the late eighteenth century. For a period of some 200 years, both theory and methodology were predominantly developed in the classical frequency domain. Around the 1960s and in tandem with the rapid development of aerospace technology, the 'state space' approach emerged as a more accurate and more powerful tool to tackle control theory study and control systems design. The concept of 'states' was introduced in order to describe the essential properties of a control system. These system states are, however, internal quantities of a control system; and therefore, in many cases they might not be directly accessible, or even faithfully measurable due to unreliability and/or high cost of sensors.

Specific techniques are, therefore, required to estimate the states, in order to design control schemes based on the system states. These techniques are called 'state estimators' or 'filters'. The most well-known filter is the Kalman filter, which was developed by Rudolf Kalman and others half a century ago. The Kalman filter uses the system dynamic model and real-time input/output measurements to generate an estimation of the system states under the influence of external noises. It is an efficient and effective estimation approach for systems subject to statistical noises. However, a limit of the Kalman filter is that it is only applicable to linear control systems. Meanwhile, almost all control systems in the real world unfortunately display nonlinearities, and are, therefore, nonlinear systems. In the real world, therefore, nonlinear state estimation becomes essential for all practical control systems.

This book addresses the issue of state estimation for nonlinear control systems. The book starts with an introduction to dynamic control systems and system states, as well as a brief description of the Kalman filter. In the following chapters, various state estimation techniques for nonlinear systems are discussed. Some of the most common methods covered in relation to nonlinear state estimation are the Extended Kalman Filters (EKFs). However, for an EKF design, an accurate plant model and Jacobians of the plant and measurement dynamic models are required. In the past two decades, Jacobian-free filters such as Unscented Kalman Filters (UKFs) and Cubature Kalman Filters (CKFs) have been investigated to circumvent some of the inherent, and adverse, issues of the EKF. This book will mainly focus on these Jacobian-free filters, and especially on the Cubature Kalman Filter (CKFs) and its extensions. Several extensions to the CKF will be derived, including Cubature Information Filters (CIF) and cubature $\mathcal{H}_\infty$ filters (C$\mathcal{H}_\infty$F). The CIF is an extension of CKF in the information domain, and it is suitable for nonlinear state estimation with multiple sensors. For nonlinear systems with non-Gaussian noises, the C$\mathcal{H}_\infty$F is a more effective candidate, and is developed from CKF and $\mathcal{H}_\infty$ filter. The square-root versions of the CIF and the C$\mathcal{H}_\infty$F will also be further derived for enhanced numerical stability. A number of case studies are also presented in the book to demonstrate the applicability of these Jacobian-free filtering approaches for nonlinear systems.

This book is primarily intended for control systems researchers, who are interested in nonlinear control systems. However, it can also be readily used as a textbook for postgraduate or senior undergraduate students on nonlinear control system courses. Lastly, practising control engineers might find this book useful in relation to the design of nonlinear control systems using state feedback techniques. Good prerequisites for making the most of this book are a sound knowledge of classical control and state variable control courses at undergraduate level, as well as (at least some elementary) knowledge of random signals and stochastic processes.

The authors are grateful to many of their current and past colleagues in preparing this manuscript. The long list includes, but is by no means limited to, Prof. Ian Postlethwaite, Dr. Naeem Khan, Dr. Sajjad Fekriasl, Prof. Christopher Edwards, Dr. Mangal Kothari, Dr. Rosli Omar, Dr. Rihan Ahmed, Dr. Devendra Potnuru, Dr. Bijnan Bandyopadhyay and Dr. Ienkaran Arasaratnam. As always, the assistance received from Springer Editors and comments from the anonymous book reviewers are equally highly appreciated by the authors.

Leicester, UK
February 2018

Kumar Pakki Bharani Chandra
Da-Wei Gu

Contents

1 Control Systems and State Estimation 1
1.1 Introductory Remarks 1
1.2 Linear and Nonlinear Control Systems 2
1.3 Control System Design and System States 4
1.4 Kalman Filter and Further Developments 6
1.5 What is in This Book 7
References 9

2 State Observation and Estimation 13
2.1 Introductory Remarks 13
2.2 Mathematical Preliminary 14
2.3 Desired Properties of State Estimators 21
2.4 Least Square Estimator and Luenberger State Observer 22
2.5 Luenberger State Observer for a DC Motor 25
2.6 Summary 27
References 27

3 Kalman Filter and Linear State Estimations 29
3.1 Introductory Remarks 29
3.2 The Discrete-Time Kalman Filter 30
3.2.1 Process and Measurement Models 30
3.2.2 Derivation of the Kalman Filter 31
3.3 Kalman Information Filter 34
3.4 Discrete-Time $\mathcal{H}_\infty$ Filter 36
3.5 State Estimation and Control of a Quadruple-Tank System 39
3.5.1 Quadruple-Tank System 39
3.5.2 Sliding Mode Control of the Quadruple-Tank System 43
3.5.3 Combined Schemes: Simulations and Results 47
3.6 Summary 55
References 56

4 Jacobian-Based Nonlinear State Estimation . . . 59
4.1 Introductory Remarks . . . 59
4.2 Extended Kalman Filter . . . 59
4.3 Extended Information Filter . . . 61
4.4 Extended $\mathcal{H}_\infty$ Filter . . . 64
4.5 A DC Motor Case Study . . . 65
4.6 Nonlinear Transformation and the Effects of Linearisation . . . 67
4.7 Summary . . . 72
References . . . 73

5 Cubature Kalman Filter . . . 75
5.1 Introductory Remarks . . . 75
5.2 CKF Theory . . . 75
5.2.1 Prediction . . . 76
5.2.2 Measurement Update . . . 76
5.2.3 Cubature Rules . . . 77
5.3 Cubature Transformation . . . 79
5.3.1 Polar to Cartesian Coordinate Transformation—Cubature Transformation . . . 81
5.4 Study on a Brushless DC Motor . . . 83
5.4.1 BLDC Motor Dynamics . . . 84
5.4.2 Back EMFs and PID Controller . . . 85
5.4.3 Initialisation of the State Estimators . . . 89
5.4.4 BLDC Motor Experiments . . . 89
5.5 Summary . . . 95
References . . . 96

6 Variants of Cubature Kalman Filter . . . 97
6.1 Introductory Remarks . . . 97
6.2 Cubature Information Filters . . . 97
6.2.1 Information Filters . . . 98
6.2.2 Extended Information Filter . . . 99
6.2.3 Cubature Information Filter . . . 100
6.2.4 CIF in Multi-sensor State Estimation . . . 102
6.3 Cubature $\mathcal{H}_\infty$ Filters . . . 104
6.3.1 $\mathcal{H}_\infty$ Filters . . . 104
6.3.2 Extended $\mathcal{H}_\infty$ Information Filter . . . 105
6.3.3 Cubature $\mathcal{H}_\infty$ Filter . . . 107
6.3.4 Cubature $\mathcal{H}_\infty$ Information Filter . . . 109
6.4 Square-Root Version Filters . . . 112
6.4.1 Square-Root Extended Kalman Filter . . . 112
6.4.2 Square-Root Extended Information Filter . . . 113
6.4.3 Square-Root Extended $\mathcal{H}_\infty$ Filter . . . 114
6.4.4 Square-Root Cubature Kalman Filter . . . 116

6.4.5 Square-Root Cubature Information Filter 118
6.4.6 Square-Root Cubature $\mathcal{H}_\infty$ Filter 120
6.4.7 Square-Root Cubature $\mathcal{H}_\infty$ Information Filter 124
6.5 State Estimation of a Permanent Magnet Synchronous Motor 129
6.5.1 PMSM Model 129
6.5.2 State Estimation Using SREIF and SRCIF 130
6.5.3 State Estimation Using SRCKF and SRC$\mathcal{H}_\infty$F 132
6.5.4 State Estimation with Multi-sensors Using SREIF, SRCIF and SRC$\mathcal{H}_\infty$IF 135
6.6 State Estimation of a Continuous Stirred Tank Reactor (CSTR) 143
6.6.1 The Model 143
6.6.2 State Estimation in the Presence of Non-Gaussian Noises 144
6.6.3 State Estimation with Near-Perfect Measurements 145
6.7 Summary 147
References 147

7 More Estimation Methods and Beyond 149
7.1 Introductory Remarks 149
7.2 Unscented Kalman Filter 149
7.2.1 Unscented Transformation 150
7.2.2 Unscented Kalman Filter 153
7.3 State-Dependent Riccati Equation (SDRE) Observers 155
7.4 SDRE Information Filter 157
7.4.1 SDREIF in Multi-sensor State Estimation 158
7.5 PMSM Case Revisited 160
7.6 Filter Robustness Consideration 163
7.6.1 Uncertainties and Robustness Requirement 163
7.6.2 Compensation of Missing Sensory Data 164
7.6.3 Selection of Linear Prediction Coefficients and Orders 168
7.6.4 A Case Study 172
7.7 Summary 180
References 180

Index 183

List of Figures

Fig. 1.1	A box furnace heating plant	2
Fig. 1.2	An example of a combined state estimation-control approach	5
Fig. 1.3	Flow chart of the book	8
Fig. 2.1	Open-loop block diagram	14
Fig. 2.2	Normal probability density function	19
Fig. 2.3	Luenberger (full-order) state observer	24
Fig. 2.4	States of Luneberger observer for dc motor	26
Fig. 2.5	Luneberger observer estimation error for dc motor	27
Fig. 3.1	A combined SMC-$\mathcal{H}_\infty$ filter approach	40
Fig. 3.2	Quadruple-Tank System (Johansson 2000)	41
Fig. 3.3	Proposed scheme using SMC and $\mathcal{H}_\infty$ filter for quadruple-tank system	48
Fig. 3.4	Actual and estimated states of the quadruple-tank using the Kalman filter in the presence of Gaussian noises	49
Fig. 3.5	Actual and estimated states of the quadruple-tank using $\mathcal{H}_\infty$ filter in the presence of Gaussian noises	50
Fig. 3.6	Estimation errors for the quadruple-tank using Kalman and $\mathcal{H}_\infty$ filters in the presence of Gaussian noises	51
Fig. 3.7	RMSEs using Kalman and $\mathcal{H}_\infty$ filters in the presence of non-Gaussian noises	51
Fig. 3.8	Actual and estimated states of the quadruple-tank using the Kalman filter in the presence of non-Gaussian noises	52
Fig. 3.9	Actual and estimated states of the quadruple-tank using $\mathcal{H}_\infty$ filter in the presence of non-Gaussian noises	53
Fig. 3.10	Estimation errors for the quadruple-tank using Kalman and $\mathcal{H}_\infty$ filters in the presence of non-Gaussian noises	54
Fig. 3.11	RMSEs using Kalman and $\mathcal{H}_\infty$ filters in the presence of non-Gaussian noises	54
Fig. 4.1	Actual and estimated velocity using EKF	66
Fig. 4.2	Estimation error of EKF	67

Fig. 4.3 Actual and estimated velocity using E$\mathcal{H}_\infty$F 67
Fig. 4.4 Estimation error using E$\mathcal{H}_\infty$F 68
Fig. 4.5 2000 random points (∗) are generated with range and bearings, which are uniformly distributed between ±0.02 and ±20° 70
Fig. 4.6 2000 random measurements are generated with range and bearings, which are uniformly distributed between ±0.02 and ±20°. These random points are then processed through the nonlinear polar to cartesian coordinate transformation and are shown as ∗. The true mean and the uncertainty ellipse are represented by • and solid line, respectively 71
Fig. 4.7 2000 random measurements are generated with range and bearings, which are uniformly distributed between ±0.02 and ±20°. These random points are then processed through the nonlinear polar to cartesian coordinate transformation and are shown as ∗. The true mean and the linearised mean are represented by • and ♦, and true and linearised uncertainty ellipses are represented by solid and dotted lines, respectively 72
Fig. 5.1 2000 random measurements are generated with range and bearings, which are uniformly distributed between ±0.02 and ±20°. These random points are then processed through the nonlinear polar to cartesian coordinate transformation and are shown as ∗. The true-, linearised-, and cubature transformation means are represented by •, ♦, and ⋆, respectively. True, linearised, and cubature transformation uncertainty ellipses are represented by solid, dotted, dashed-dotted and dashed lines, respectively 82
Fig. 5.2 Block diagram of full state feedback BLDC motor 83
Fig. 5.3 Proposed scheme for a BLDC motor 88
Fig. 5.4 Experimental setup for BLDC drive 90
Fig. 5.5 Block diagram for BLDC drive experiment 90
Fig. 5.6 Speed controller scheme using CKF 91
Fig. 5.7 Experimental results for low reference speed (30 rpm) 92
Fig. 5.8 Experimental results for high reference speed (2000 rpm) 93
Fig. 5.9 Experimental results with load variations 94
Fig. 5.10 Experimental results with stator resistance variations 94
Fig. 5.11 Estimation errors with de-tuned stator inductance 95
Fig. 6.1 Actual and estimated states of PMSM using SREIF and SRCIF 131
Fig. 6.2 RMSE of PMSM using SREIF and SRCIF 131
Fig. 6.3 Actual and estimated states using decentralised SREIF and SRCIF 132
Fig. 6.4 RMSE of PMSM using decentralised SREIF and SRCIF 133

Fig. 6.5 Actual and estimated states of PMSM for non-Gaussian noise using SRCKF and SRC$\mathcal{H}_\infty$F ... 134
Fig. 6.6 RMSEs of the PMSM using SRCKF and SRC$\mathcal{H}_\infty$F ... 134
Fig. 6.7 RMSEs of x_3 with near-perfect measurements for the PMSM using SRCKF and SRC$\mathcal{H}_\infty$F ... 135
Fig. 6.8 PMSM states using decentralised SRE$\mathcal{H}_\infty$IF and decentralised SRC$\mathcal{H}_\infty$IF for low Gaussian noise ... 136
Fig. 6.9 RMSE of PMSM speed using decentralised SRE$\mathcal{H}_\infty$IF and decentralised SRC$\mathcal{H}_\infty$IF for low Gaussian noise ... 137
Fig. 6.10 PMSM states using decentralised SRE$\mathcal{H}_\infty$IF and decentralised SRC$\mathcal{H}_\infty$IF for high Gaussian noise ... 138
Fig. 6.11 RMSE of PMSM speed using decentralised SRE$\mathcal{H}_\infty$IF and decentralised SRC$\mathcal{H}_\infty$IF for high Gaussian noise ... 138
Fig. 6.12 PMSM states using decentralised SRE$\mathcal{H}_\infty$IF and decentralised SRC$\mathcal{H}_\infty$IF for low non-Gaussian noise ... 139
Fig. 6.13 RMSE of PMSM speed using decentralised SRE$\mathcal{H}_\infty$IF and decentralised SRC$\mathcal{H}_\infty$IF for low non-Gaussian noise ... 140
Fig. 6.14 PMSM states using decentralised SRE$\mathcal{H}_\infty$IF and decentralised SRC$\mathcal{H}_\infty$IF for high non-Gaussian noise ... 141
Fig. 6.15 RMSE of PMSM speed using decentralised SRE$\mathcal{H}_\infty$IF and decentralised SRC$\mathcal{H}_\infty$IF for high non-Gaussian noise ... 142
Fig. 6.16 PMSM simulation results in the presence of near-perfect Gaussian and non-Gaussian noise in multi-sensor SRC$\mathcal{H}_\infty$IF ... 142
Fig. 6.17 Actual and estimated states of CSTR for non-Gaussian noise using SRCKF and SRC$\mathcal{H}_\infty$F ... 145
Fig. 6.18 CSTR's RMSEs of x_1 in the presence of non-Gaussian noise using SRCKF and SRC$\mathcal{H}_\infty$F ... 146
Fig. 6.19 CSTR's RMSEs of x_1 in the presence of near-perfect measurement using SRCKF and SRC$\mathcal{H}_\infty$F ... 146
Fig. 7.1 2000 random measurements are generated with range and bearings, which are uniformly distributed between ± 0.02 and $\pm 20°$. These random points are then processed through the nonlinear polar to cartesian coordinate transformation and are shown as $*$. The true, linearised and unscented transformation means are represented by •, ♦ and ■, respectively. True, linearised and unscented transformation uncertainty ellipses are represented by solid, dotted and dashed-dotted lines, respectively ... 153
Fig. 7.2 Actual and measured data ... 162
Fig. 7.3 State estimation using single- and multi-sensor SDREIF ... 162
Fig. 7.4 Error plots using single- and multi-sensor SDREIF ... 163
Fig. 7.5 MSD two cart system ... 173
Fig. 7.6 Performance of KF without measurement loss ... 175

Fig. 7.7 KF without measurement loss for states x_1, x_3 and x_4 175
Fig. 7.8 Estimation of state x_2 176
Fig. 7.9 Estimation analysis of states x_1, x_3 and x_4 177
Fig. 7.10 Estimation error of state x_2 178
Fig. 7.11 Estimation error of states x_1, x_3 and x_4 178
Fig. 7.12 Kalman filter gains of ZeroGKF approach 179
Fig. 7.13 Kalman filter gains of ComKF approach 179

List of Tables

Table 1.1 Key differences between Wiener and Kalman filters (Anderson and Moore 1979) . . . 6
Table 3.1 Quadruple-tank parameters. . . . 42
Table 5.1 Back EMFs as functions of θ and ω . . . 84
Table 5.2 The functions f_{ab}, f_{bc} and f_{ca} . . . 86
Table 5.3 Switching functions for hysteresis current control . . . 86

List of Algorithms

Algorithm 3.1	The Kalman filter (KF)	35
Algorithm 3.2	$\mathcal{H}_\infty$ Filter	39
Algorithm 4.1	Extended Kalman filter	61
Algorithm 4.2	Extended Information Filter	63
Algorithm 4.3	Extended $\mathcal{H}_\infty$ filter	65
Algorithm 5.1	Cubature Kalman filter	79
Algorithm 5.2	Cubature Transform	81
Algorithm 6.1	Cubature information filter	103
Algorithm 6.2	Cubature $\mathcal{H}_\infty$ filter	108
Algorithm 6.3	Cubature $\mathcal{H}_\infty$ information filter	110
Algorithm 6.4	Square-root extended Kalman filter	113
Algorithm 6.5	Square-root cubature information filter	119
Algorithm 6.6	Square-root cubature $\mathcal{H}_\infty$ filter	123
Algorithm 6.7	Square-root cubature $\mathcal{H}_\infty$ information filter	127
Algorithm 7.1	Unscented Transform	150
Algorithm 7.2	Unscented Kalman Filter	154
Algorithm 7.3	State-Dependent Riccati Equation Information Filter (SDREIF)	159
Algorithm 7.4	Selection of the LP filter order (first method)	171
Algorithm 7.5	Selection of the LP filter order (second method)	172

Chapter 1
Control Systems and State Estimation

1.1 Introductory Remarks

This book is about state estimation of control systems, nonlinear control systems in particular. Control systems are dynamic systems and exist in engineering, physical sciences as well as in social sciences. The earliest, somehow commonly recognised control system could be traced back to James Watt's flyball governor in 1769. The study on control systems, analysis and system design, has been continuously developed ever since. Until the mid of last century the control systems under investigation had been single-input-single-output (SISO), time-invariant systems and were mainly deterministic and of lumped parameters. The approaches used were of frequency domain nature, so-called classical approaches. In classical control approaches, control system's dynamic behaviour is represented by transfer functions. Rapid developments and needs in aerospace engineering in the 1950 and 1960s greatly drove the development of control system theory and design methodology, particularly in the state-space approach that is powerful towards multi-input-multi-output (MIMO) systems. State-space models are used to describe the dynamic changes of the control system. In the state-space approach, the system states determine how the system is dynamically evolving. Therefore, by knowing the system states one can know the system's properties and by changing the trajectory of system states successfully with specifically designed controllers one can achieve the objectives for a certain control system.

However, the states of a control system are "internal" parameters and are not always accessible. Figure 1.1 is a sketch diagram of a box furnace heating plant. Parts to be heated are placed within the furnace and heated by the gas and air (the input) pumped in from one end of the furnace. The control objective is to control the temperature and the heating time of the part by adjustment of the input. In this system, the temperature within the furnace is the state variable. Strictly speaking this plant is a distributive control system. By partitioning, however, it can be represented by lumped parameter model satisfactorily. Even so, it is difficult to measure the temperature of the part in the furnace due to the high temperature which requires

K. P. B. Chandra and D.-W. Gu, *Nonlinear Filtering*,
https://doi.org/10.1007/978-3-030-01797-2_1

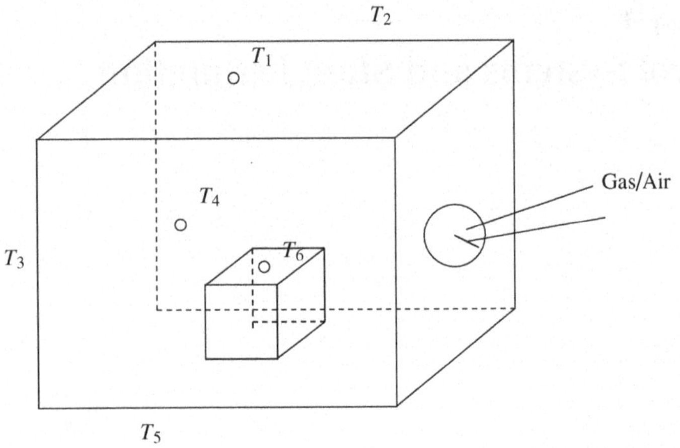

Fig. 1.1 A box furnace heating plant

expensive thermocouple (sensor) of top quality. That may not be a solution the Industry would accept. State estimation methods are necessary in such a scenario.

Another major reason for the need of state estimation is due to the existence of noises and plant uncertainties. Industrial control systems are always operating in real environments and are inevitably affected by random noises and/or disturbances. On the other hand, the mathematical representation of the system dynamics, namely, the system model, is difficult to be derived fully and accurately, due to the complexity of the plant and cost in modelling and controller design/operation. The initial condition of the operation of the control system is usually unknown. Even if a system model is accurate initially, the model parameters would vary due to tear-and-wear. Hence, the idea of using the model and input to work out the system states is impractical. State estimation methods are indeed needed in practice. In the following sections of this chapter, more control system properties will be introduced and the use of state estimation is to be further justified.

1.2 Linear and Nonlinear Control Systems

A control system or plant or process is an interconnection of components to perform certain tasks and to yield a desired response, i.e. to generate desired signals (the output), when it is driven by manipulating (a group of) signals (the input). In this book, a control system is a causal, dynamic system, i.e. the output depends on not only the present input but also the input applied at the previous time.

Due to the increasing complexity of physical systems to be controlled and rising demands on system properties, most industrial control systems are no longer single-input and single-output (SISO) but multi-input and multi-output (MIMO) systems with a high interrelationship (coupling) between these channels. The number of (state) variables in a system could be very large as well. These systems are called multivariable systems.

In order to analyse and design a control system, it is advantageous if a mathematical representation of such a relationship (namely, a model) is available. The system dynamics is usually governed by a set of differential equations, either ordinary differential equations, for lumped parameter systems which are the systems this book considers, or partial differential equations for distributive systems.

By introducing a set of appropriate state variables, the dynamic behaviour of a time-invariant, control system may be represented by the following mathematical equations (the model), in the continuous-time case:

$$\begin{aligned}\dot{\mathbf{x}}(t) &= \mathbf{f}(\mathbf{x}(t), \mathbf{u}(t)) + \mathbf{w}(t)\\ \mathbf{z}(t) &= \mathbf{h}(\mathbf{x}(t), \mathbf{u}(t)) + \mathbf{v}(t)\end{aligned} \tag{1.1}$$

where $\mathbf{x}(t) \in \mathbf{R}^n$ is the state vector, $\mathbf{u}(t) \in \mathbf{R}^m$ the input (control) vector, and $\mathbf{z}(t) \in \mathbf{R}^p$ the output (measurement) vector; and $\mathbf{w}(t)$ and $\mathbf{v}(t)$ are process noise and measurement noise, respectively.

In the discrete-time case, the model is as

$$\mathbf{x}_k = \mathbf{f}(\mathbf{x}_{k-1}, \mathbf{u}_{k-1}) + \mathbf{w}_{k-1} \tag{1.2}$$

$$\mathbf{z}_k = \mathbf{h}(\mathbf{x}_k, \mathbf{u}_k) + \mathbf{v}_k \tag{1.3}$$

where k is the time index.

The noises $\mathbf{w}$ and $\mathbf{v}$ are assumed to be zero mean Gaussian-distributed, random variables with covariances of $\mathbf{Q}_{k-1}$ and $\mathbf{R}_k$.

The continuous-time and discrete-time models listed above are generic. They may represent just the plant dynamics in the open-loop system scenario or the dynamics of both the plant and controller in the closed-loop system scenario.

In this book, control systems are mainly considered in their discrete-time form.

Control systems can be classified as linear systems or nonlinear systems. A control system is linearif the functions $\mathbf{f}$ and $\mathbf{h}$ in (1.1) and (1.1) satisfy the superposition rules with regard to the state $\mathbf{x}$, the input $\mathbf{u}$ as well as the initial state $\mathbf{x}(0)$.

In the linear case , the system model can be described as

$$\begin{aligned}\dot{\mathbf{x}}(t) &= A\mathbf{x}(t) + B\mathbf{u}(t) + \mathbf{w}(t)\\ \mathbf{y}(t) &= C\mathbf{x}(t) + D\mathbf{u}(t) + \mathbf{v}(t)\end{aligned} \tag{1.4}$$

In the case of discrete-time systems, similarly the model is given by

$$\mathbf{x}_k = \mathbf{F}_{k-1}\mathbf{x}_{k-1} + \mathbf{G}_{k-1}\mathbf{u}_{k-1} + \mathbf{w}_{k-1}, \tag{1.5}$$

$$\mathbf{z}_k = \mathbf{H}_k\mathbf{x}_k + \mathbf{v}_k, \tag{1.6}$$

1.3 Control System Design and System States

Control systems design is necessary for almost all practical systems. The task of a control engineer is to design a controller to stabilise an unstable system and/or to achieve the desired performance in the presence of uncertainties, including dynamic perturbations (parametric variations and unmodelled dynamics), and disturbance and noises. In general, there are two categories of control systems, the open-loop and closed-loop control systems (Gu et al. 2014). Open-loop systems are "simple" in structure and less costly. In an open-loop control system, the output has no effect on the control action. For a given input, the system gives a certain output, if in an uncertainty-free scenario. In the presence of dynamic perturbations, the stability of open-loop systems is not guaranteed, even if the originally built, nominal system is stable. Similarly, the tracking performance would be affected by the disturbance in an open-loop structure. In open-loop systems, no measurements are made at the output for control purpose and hence it does not have the feedback mechanism. In contrast to an open-loop system, a closed-loop control system uses sensors/transducers to measure the actual output to adjust the input in order to achieve desired output. The measure of the output in this use is called the feedback signal, and a closed-loop system is also called a feedback system. In a closed-loop system, the feedback signals are compared with the reference signals, which possibly produces errors in most cases. Based on these error signals, the controller generates the inputs to the system which help the outputs to reach their desirable values.

Any dynamical system can be characterised by a set of state variables. If the state variables are used to obtain the control signals, it is called as a state feedback control system and if the controller is directly based on the measured outputs, then it is called as output feedback systems. In most of the real-world applications it is not always possible to access the complete state information due to the limitations on sensors and/or cost consideration, such as in the box furnace heating system introduced earlier. If *some* of the sensors are very noisy or expensive or physically unsuitable for the control system, then it is not advisable to use them to measure the states either. Rather, one can use the state estimation methods to obtain or, more accurately, to estimate the unavailable states using the input and output information together with known dynamic structure of the system.

Control systems can be classified in several ways like: linear and nonlinear, deterministic and stochastic, time-varying and time-invariant, lumped parameter and distributive parameter systems, etc. Control schemes for these systems can be designed based on the output or state information. State feedback controllers include pole placement control, linear quadratic regulator (LQR), dynamic inversion control, etc. and output feedback controllers include proportional-integral-derivative (PID) control, linear quadratic Gaussian control, etc. Many controllers can be designed based on either state feedback or output feedback like sliding mode control, $\mathcal{H}_\infty$ control, model predictive control, etc. One would also note that even if the complete state information is not available, the state feedback controllers could still be designed using estimated states obtained from state estimation methods. A combined state

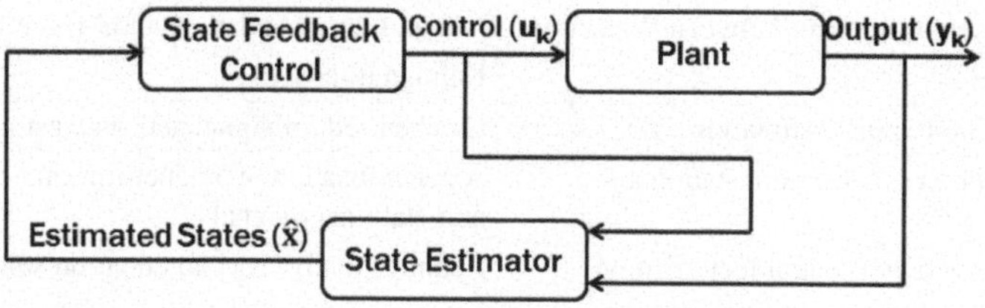

Fig. 1.2 An example of a combined state estimation-control approach

estimation-control approach is shown in Fig. 1.2. If the state estimator in Fig. 1.2 is the Kalman filter, then this approach may lead to a linear quadratic Gaussian (LQG) controller. A similar kind of approach is to be explored later in the book using a sliding mode control and an $\mathcal{H}_\infty$ filter on a quadruple-tank system. The use of Kalman filter and $\mathcal{H}_\infty$ filter will be compared via simulations.

State estimation is a process of estimating the unmeasured states using the noisy outputs, control inputs along with process and measurement models. It has been an active research area for several decades. Similar to control systems, state estimation can also be classified as linear and nonlinear, deterministic and stochastic, etc. The earliest state estimation problem was considered in the field of astronomical studies by *Karl Friedrich Gauss* in 1795, where the planet and comet motion was studied using the telescopic measurements (Sorenson 1970). Gauss used the least square method as the estimation tool. After more than 140 years of Gauss' invention, *Andrey Nikolaevich Kolmogorov* (Kolmogorov 2018) and *Norbert Wiener* (Wiener 1949) solved the linear least square estimation problem for stochastic systems. Kolmogorov studied discrete least-estimation problems, whereas, Wiener studied the continuous-time problems (Kailath 1968). Wiener filter[1] is also a useful tool in signal processing and communication theory. But when it specifically comes to the state estimation, Wiener filter is seldom used as it only deals with the stationary processes. *Rudolf Emil Kalman* extended Wiener's work for more generic non-stationary processes in the path breaking paper (Kalman 1960). The Wiener filter was developed in the frequency domain and is mainly used for signal estimation, whereas, the Kalman filter was developed in the time domain for state estimation. Key differences between Wiener and Kalman filters can be given in Table 1.1.

As the main emphasis of this book is on the state estimation, the Wiener filter will not be further discussed.

[1]In general, the term "filter" is frequently used for state estimators in the estimation literature. This is due to Wiener, who studied the continuous-time estimation problem and noted that his algorithm can be implemented using a linear circuit. In circuit theory, the filters are used to separate the signals over different frequency ranges. Wiener's solution extended the classical theory of filter design to problems of obtaining the filtered signals from noisy measurements (Kailath et al. 2000).

Table 1.1 Key differences between Wiener and Kalman filters (Anderson and Moore 1979)

Wiener filter	Kalman filter
Mainly used for signal estimation	Can be used for signal and state estimation
Signal and process noises are stationary	A generalisation of Wiener filter for non-stationary signals
Can be obtained by spectral factorisation methods	Requires matrix Riccati equation solutions
Basically, a frequency domain approach	A time domain (state-space) approach

1.4 Kalman Filter and Further Developments

The Kalman filter can be defined as "*an estimator used to estimate the state of a linear dynamic system perturbed by Gaussian white noise using measurements that are linear functions of the system state but corrupted by additive Gaussian white noise*" (Grewal and Andrews 2001). The Kalman filter and its variants are the main estimation tools for practical systems in the past several decades. The Kalman filter can be represented in an alternative form as the *information filter*, where the parameters of interest are the information states and the inverse of covariance matrix rather than the states and covariance matrix. Information filters are more flexible in initialisation compared to conventional Kalman filters and the update stage is computationally economic, and it can be readily extended for multi-sensor fusion; for more details please see Anderson and Moore (1979), Mutambara (1998), and will be discussed further later in the book. Both Kalman and information filters can be derived in the Gaussian framework and they need accurate process and measurement models.

The early success of the Kalman filter in 1960s in aerospace applications was due to the availability of accurate system models, which were obtained after spending millions of dollars on the space programmes (Simon 2006). However, it is not worth to spend that huge amount of money in most other industrial applications in order to get an accurate model. One of the alternatives to the Kalman filter is to develop the estimator using the concepts of robust control. Several researchers have explored the robust control theory, specially the $\mathcal{H}_\infty$ theory, to develop robust state estimators (Grimble and Sayed 1990; Napgal and Khargonekar 1991; Yaesh and Shaked 1992; Shaked 1990; Basar 1991; Banavar and Speyer 1991). In $\mathcal{H}_\infty$ filters, the requirements on the accurate models or 'apriori' statistical noise properties can be relaxed to a certain extent.

In real-time implementation of Kalman filters, the propagated error covariance matrices may become ill-conditioned, which may eventually hinder the filtering operation. This can happen if some of the states are measured with greater precision than others, where the elements of covariance matrix corresponding to accurately measured states will have lower values, while the other entries will have higher values. These types of ill-conditioned covariance matrices may cause numerical instability during the online implementation. To circumvent these difficulties, one can use square-root Kalman filters, where the square root of the error covariance matrices are propagated. Key properties of the square-root filters include symmetric positive definiteness of error covariances, availability of square-root factors, doubled order precision, improved numerical accuracy, etc. Kailath et al. (2000), Kaminski et al. (1971), Morf and Kailath (1975), Verhaegen and Van Dooren (1986), Carlson (1990), which are useful in the filter implementation and applications. Similar to square-root Kalman filters, the information and $\mathcal{H}_\infty$ filters can also be explored as square-root information filters (Bierman et al. 1990; Psiaki 1999; Park and Kailath 1995) and square-root $\mathcal{H}_\infty$ filters (Kailath et al. 2000; Hassibi 2000).

Initially the Kalman filter was developed for linear systems. However, most of the real-world problems are nonlinear and hence the Kalman filter has been further extended for nonlinear systems. *Stanley F. Schmidt* was the first researcher to explore the Kalman filter for nonlinear systems, while doing so he developed the so-called *extended Kalman filter* (EKF), see McGee and Schmidt (1985) for fascinating historical facts about the development of the EKF for practical applications. However, EKFs are only suitable for "mild" nonlinearities where the first-order approximations of the nonlinear functions are available and EKFs also require evaluation of state Jacobians at every iterations. To overcome some of the limitations of EKF, an unscented Kalman filter (UKF) has been proposed Julier and Uhlmann (2004), Julier et al. (2000), which is a derivative-free filter. The UKF uses the deterministic sampling approach to capture the mean and covariances with sigma points and in general has been shown to perform better than EKF in nonlinear state estimation problems. The UKF has been further explored in the information domain for decentralised estimation (Vercauteren and Wang 2005; Campbell and Whitacre 2007; Lee 2008). There are a few other nonlinear estimation techniques found in the literature, to name a few, Rao-Blackwellised particle filters (Doucet et al. 2000), which are the improved version of particle filters (Arulampalam et al. 2002), Gaussian filters (Ito and Xiong 2000), state-dependent Riccati equation filters (Mracek et al. 1996; Nemra and Aouf 2010), sliding mode observers Xiong and Saif (2001), Fourier-Hermite Kalman filter Sarmavuori and Sarkka (2012), etc.

Recently, the cubature Kalman filter (CKF) (Arasaratnam and Haykin 2009) has been proposed for nonlinear state estimation. CKF is a Gaussian approximation of Bayesian filter, but provides a more accurate filtering estimate than existing Gaussian filters. In this book, the CKF will be further explored for multi-sensor state estimation and for non-Gaussian noises. The efficacy of the proposed methods are demonstrated on various simulation examples.

1.5 What is in This Book

A brief introduction can be found in the first two chapters on control systems, system states and state estimation. In addition, a DC motor is used to illustrate the usage of the classical Luenberger state observer in Chap. 2. This book will then introduce

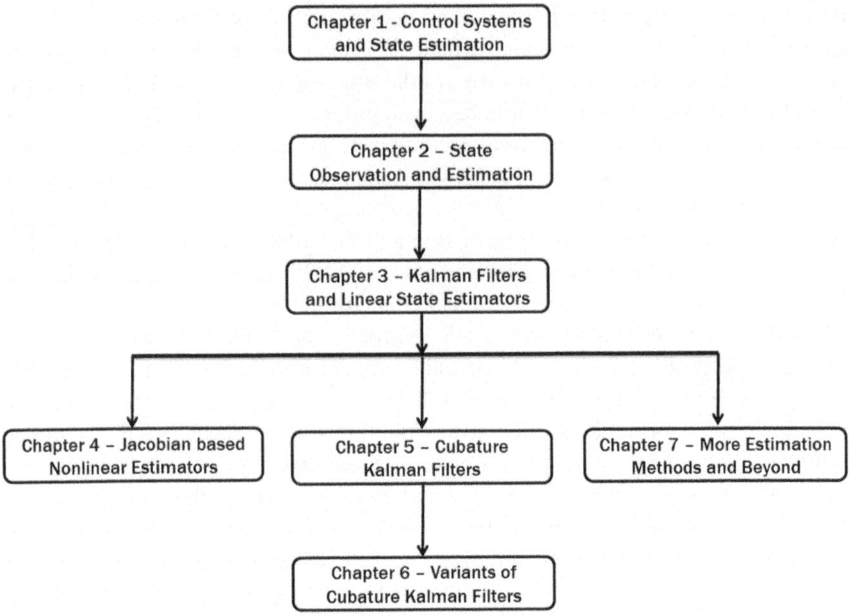

Fig. 1.3 Flow chart of the book

and discuss several important state estimation approaches together with application of these approaches in a few case studies. A graphical representation of the book is shown in Fig. 1.3.

Before discussions on state estimation for nonlinear control systems that is the main theme of the book, Chap. 3 introduces the linear estimation techniques first. It begins with mathematical preliminaries of Kalman filter, information filter and $\mathcal{H}_\infty$ filter. The detailed derivation of Kalman filter and the game theory approach to the discrete $\mathcal{H}_\infty$ filter is briefly discussed. Towards the end of this chapter, a combined use of the sliding mode control (SMC) and an $\mathcal{H}_\infty$ filter for a quadruple-tank system is to be proposed. It is assumed that, out of the system's four states only two states are directly accessible. The complete state vector is estimated using an $\mathcal{H}_\infty$ filter and the SMC is designed based on the estimated states. The proposed $\mathcal{H}_\infty$ filter based SMC can be easily extended for other practical systems.

From Chap. 4 the book concentrates on nonlinear state estimation techniques. Chapter 4 presents several Jacobian-based nonlinear state estimation approaches, including the extended Kalman filter (EKF), extended information filter (EIF) and extended $\mathcal{H}_\infty$ filter (E$\mathcal{H}_\infty$F). These approaches all have a linear case origin and are corresponding extensions for nonlinear control system state estimations.

Chapter 5 focusses on the cubature Kalman filter (CKF), the idea behind this approach and its derivation based on cubature transformation. CKF is a Gaussian approximation of Baysian filtering, but provides a more accurate estimation than other existing Gaussian filters.

Chapter 6 presents variants of cubature Kalman filters, including the cubature information filter (CIF), the cubature $\mathcal{H}_\infty$ filter (C$\mathcal{H}_\infty$F) and the cubature $\mathcal{H}_\infty$ information filter (C$\mathcal{H}_\infty$IF). CIF is the fusion of extended $\mathcal{H}_\infty$ filter and CKF. It can be easily extended to multi-sensor state estimation, where the data from various nonlinear sensors are fused. The C$\mathcal{H}_\infty$F is derived for state estimation of nonlinear systems with general noises; not limited to Gaussian noises. For numerical accuracy, the square-root versions of introduced nonlinear filtering approaches introduced so far are presented. Among that, the square root cubature information filter (SRCIF) is developed using the J-unitary transformation for the single sensor as well as multi-sensor cases. The efficacy of the multi-sensor square-root CIF is validated in a permanent magnet synchronous motor example and a continuous stirred tank reactor problem, in the presence of Gaussian and non-Gaussian noises.

Chapter 7, the last chapter of the book, mainly discusses other nonlinear state estimation methods, the unscented Kalman filter (UKF) and the state-dependent Riccati equation (SDRE) observer. Chapter 7, and the book, ends at considering an important topic, the so-called "robustness" property of linear and nonlinear state estimators, which is a topic for current and future research. The considerations include inaccurate knowledge on noises and missing of measurement data. These commonly occur in real-world environment and are thus crucial for successful industrial applications of control systems state estimation approaches.

A number of application examples, including a quadruple-tank system, coordinate transformations that forms the base of simultaneous localisation and mapping (SLAM) cases study, a permanent magnet synchronous motor, a highly nonlinear and high fidelity brush-less DC motor and a continuously stirred tank reactor case, are frequently used to illustrate and to compare the use of various methods introduced in the book.

References

Anderson BDO, Moore JB (1979) Optimal filtering, vol 21. Prentice Hall, Englewood Cliffs, pp 22–95

Arasaratnam I, Haykin S (2009) Cubature kalman filters. IEEE Trans Autom Control 54(6):1254–1269

Arulampalam MS, Maskell S, Gordon N, Clapp T (2002) A tutorial on particle filters for online nonlinear/non-gaussian bayesian tracking. IEEE Trans Signal Process 50(2):174–188

Banavar RN, Speyer JL (1991) A linear-quadratic game approach to estimation and smoothing. In: Proceedings of the 1991 American control conference. IEEE, pp 2818–2822

Basar T (1991) Optimum performance levels for minimax filters, predictors and smoothers. Syst Control Lett 16(5):309–317

Bierman G, Belzer MR, Vandergraft JS, Porter DW (1990) Maximum likelihood estimation using square root information filters. IEEE Trans Autom Control 35(12):1293–1298

Campbell ME, Whitacre WW (2007) Cooperative tracking using vision measurements on seascan UAVs. IEEE Trans Control Syst Technol 15(4):613–626

Carlson NA (1990) Federated square root filter for decentralized parallel processors. IEEE Trans Aerosp Electron Syst 26(3):517–525

Doucet A, Godsill S, Andrieu C (2000) On sequential monte carlo sampling methods for bayesian filtering. Stat Comput 10(3):197–208

Grewal M, Andrews A (2001) Kalman filtering: theory and practice using matlab. Wiley, New Jersey

Grimble MJ, El Sayed A (1990) Solution of the h_∞ optimal linear filtering problem for discrete-time systems. IEEE Trans Acoust Speech Signal Process 38(7):1092–1104

Gu D-W, Petkov PH, Konstantinov MM (2014) Robust control design with MATLAB®. Springer Science & Business Media, Berlin

Hassibi B, Kailath T, Sayed AH (2000) Array algorithms for H/sup/spl infin//estimation. IEEE Trans Autom Control 45(4):702–706

Ito K, Xiong K (2000) Gaussian filters for nonlinear filtering problems. IEEE Trans Autom Control 45(5):910–927

Julier SJ, Uhlmann JK (2004) Unscented filtering and nonlinear estimation. Proc IEEE 92(3):401–422

Julier S, Uhlmann J, Durrant-Whyte HF (2000) A new method for the nonlinear transformation of means and covariances in filters and estimators. IEEE Trans Autom Control 45(3):477–482

Kailath T (1968) An innovations approach to least-squares estimation-part i: linear filtering in additive white noise. IEEE Trans Autom Control 13(6):646–655

Kailath T, Sayed AH, Hassibi B (2000) Linear estimation, vol 1. Prentice Hall, Upper Saddle River

Kalman RE (1960) A new approach to linear filtering and prediction problems. J Basic Eng 82(1):35–45

Kaminski P, Bryson A, Schmidt S (1971) Discrete square root filtering: a survey of current techniques. IEEE Trans Autom Control 16(6):727–736

Kolmogorov AN, Doyle WL, Selin I (1962) Interpolation and extrapolation of stationary random sequences. Bulletin of academic sciences, Mathematics series, USSR, vol 5, 1941. Translation by W. Doyle and J. Selin, RM-3090-PR, Rand Corporation, Santa Monica, California

Lee D-J (2008) Nonlinear estimation and multiple sensor fusion using unscented information filtering. IEEE Signal Process Lett 15:861–864

McGee LA, Schmidt SF (1985) Discovery of the kalman filter as a practical tool for aerospace and industry

Morf M, Kailath T (1975) Square-root algorithms for least-squares estimation. IEEE Trans Autom Control 20(4):487–497

Mracek CP, Clontier J, D'Souza CA (1996) A new technique for nonlinear estimation. In: Proceedings of the 1996 IEEE international conference on control applications. IEEE, pp 338–343

Mutambara AG (1998) Decentralized estimation and control for multisensor systems. CRC press, Boca Raton

Napgal K, Khargonekar P (1991) Filtering and smoothing in an h-infinity setting. IEEE Trans Autom Control 36:152–166

Nemra A, Aouf N (2010) Robust INS/GPS sensor fusion for uav localization using SDRE nonlinear filtering. IEEE Sens J 10(4):789–798

Park P, Kailath T (1995) New square-root algorithms for kalman filtering. IEEE Trans Autom Control 40(5):895–899

Psiaki ML (1999) Square-root information filtering and fixed-interval smoothing with singularities. Automatica 35(7):1323–1331

Sarmavuori J, Sarkka S (2012) Fourier-hermite kalman filter. IEEE Trans Autom Control 57(6):1511–1515

Shaked U (1990) h_∞-minimum error state estimation of linear stationary processes. IEEE Trans Autom Control 35(5):554–558

Simon D (2006) Optimal state estimation: Kalman, H infinity, and nonlinear approaches. Wiley, New Jersey

Sorenson HW (1970) Least-squares estimation: from gauss to kalman. IEEE Spectr 7(7):63–68

Vercauteren T, Wang X (2005) Decentralized sigma-point information filters for target tracking in collaborative sensor networks. IEEE Trans Signal Process 53(8):2997–3009
Verhaegen M, Van Dooren P (1986) Numerical aspects of different kalman filter implementations. IEEE Trans Autom Control 31(10):907–917
Wiener N (1949) Extrapolation, interpolation, and smoothing of stationary time series, vol 7. MIT press, Cambridge
Xiong Y, Saif M (2001) Sliding mode observer for nonlinear uncertain systems. IEEE Trans Autom Control 46(12):2012–2017
Yaesh I, Shaked U (1992) Game theory approach to optimal linear state estimation and its relation to the minimum h_∞ norm estimation. IEEE Trans Autom Control 37(6):828–831

Chapter 2
State Observation and Estimation

2.1 Introductory Remarks

Basically, the term "Filtering" is referred to a technique to extract information (signal in this context) from noise contaminated observations (measurements). If the signal and noise spectra are essentially non-overlapping, the design of a frequency domain filter that allows the desired signal to pass while attenuating the unwanted noise would be a possibility. A classical filter could be either low pass, band pass/stop or high pass. However, when the noise and information signals are overlapped in spectrum, then the design of a filter to completely separate the two signals would not be possible. In such a situation the information has to be retrieved through estimation, smoothing or prediction. Figure 2.1 below shows a general diagram of an open-loop system (plant) subject to noise contamination at the output end.

With regard to Fig. 2.1, *filtering* usually means to retrieve the (output) signal $y(t)$ from the noise contaminated data (measurement) $z(t)$ at time t, using the measurement up to the current time t (Wiener 1949). *Smoothing*, on the other hand, indicates to recover information of $y(t)$ with the help of measurements obtained later than at time t, i.e. $z(t+\tau)$ for some $\tau > 0$. It can be called a delay in producing the information of $y(t)$ as compared to the filtering operation. Another important concept is *prediction*, defined as the forecasting of information processing. In prediction, the information of $y(t+\tau)$ for some $\tau > 0$ is forecasted based on the measurement up to time t.

In the context of a control system as it being the subject of this book, the retrieval of the system state is more involved. Generally speaking, the system state is "internal" information and therefore is not directly accessible in many cases. What makes it even worse is that the system would suffer from uncertainties including dynamic perturbations, disturbances and noises. To estimate the state of a control system, it is vital to employ all available information including the system input ($u(t)$), the measurement ($z(t)$) which could be contaminated, and the system model which is usually inaccurate/incomplete though.

K. P. B. Chandra and D.-W. Gu, *Nonlinear Filtering*,
https://doi.org/10.1007/978-3-030-01797-2_2

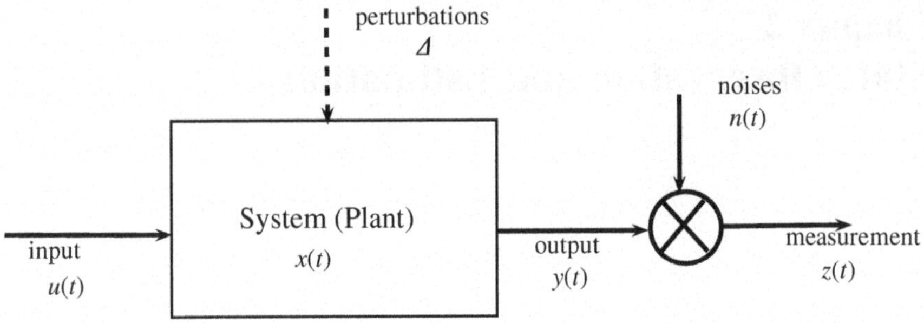

Fig. 2.1 Open-loop block diagram

State estimation has long been an active research topic in the control system engineering because of its importance in analysis of system behaviour as well as in design of control schemes such as state feedback controllers.

In this chapter, following the introduction of a few mathematical terms in the next section, a summary of desired properties of state estimation algorithms (estimators) one would expect will be presented. The chapter finishes with the simplest approach of state estimation, *namely* "state observation" for linear, time-invariant (LTI), deterministic systems, together with a DC motor case to illustrate its usage.

2.2 Mathematical Preliminary

It is useful to include in the book some basic definitions related to random variables and their statistical properties such as mean, variance, correlation and distribution of random processes. The descriptions of those are quite concise. Interested readers may find details in standard textbooks on probability and stochastic systems like Loeve (1963), Papoulis and Pillai (2002), Chung (2001).

Random Variable

In Probability Theory, a random variable is a variable whose value varies randomly, within a usually known range and subject to "chances". The possible values of a random variable might represent the possible outcomes of a yet-to-be-performed experiment. A random variable (or a random signal) cannot be characterised by a simple, well-defined mathematical function and its future values cannot be precisely predicted. However, the probability and statistical properties (e.g. *mean* and *variance*) can be used to analyse the behaviour of a random variable. If an event ε is a possible outcome of a random experiment then the probability of this event can be denoted by $p(\varepsilon)$ and generally all possible outcomes of the random experiment, if finite, represented by $\varepsilon_j, \forall j = 1, 2, \ldots n$, follow

$$0 \leq p(\varepsilon) \leq 1 \qquad \text{and} \qquad \sum_{j=1}^{n} p(\varepsilon_j) = 1 \tag{2.1}$$

The probability density function of a continuous random variable X, is a function which is summed to obtain the probability that the variable takes a value x in a given interval. It is defined as

$$f(x) = \frac{dF(x)}{dx} \tag{2.2}$$

where $F(x)$ is the cumulative distribution function.[1]

The probability density function satisfies

$$F(\infty) = \int_{-\infty}^{\infty} f(x)dx = 1 \tag{2.4}$$

Similarly, for a discrete random variable, its probability distribution (or probability function) is a list of probabilities associated with each of its possible value, i.e.

$$p(x_i) = P[X = x_i] \tag{2.5}$$

The probability distribution satisfies the criteria defined in (2.1).

Mean

The *expected* or *mean value* of a random variable indicates the average or "central" value of that random variable. The mean value gives a general idea about a random variable instead of the complete details of its probability distribution (of a discrete random variable) or the probability density function (of a continuous random variable). Two random variables with different distributions (or density functions) might have the same mean values, hence the mean value itself does not reveal the complete information of a random variable. If X is a continuous random variable with probability density function $f(x)$, the expected or mean value can be represented as

$$\mu = \mathbb{E}[X] := \int_{-\infty}^{\infty} xf(x)dx \tag{2.6}$$

Similarly, if X is a discrete random variable with possible values x_i, where $i = \{1, 2, 3 \ldots, n\}$ and its probability $p(x_i)$ is given as $P(X = x_i)$, then the mean or expected value can be defined as

---

[1]The cumulative distribution function gives the probability that the random variable X satisfies

$$F(x) = p(X < x) \qquad \forall - \infty < x < \infty \tag{2.3}$$

$$\mu = \mathbb{E}[X] := \sum x_i P(X = x_i) \tag{2.7}$$

where the elements are summed over all possible values of the random variable X.

Variance

The *variance* of a random variable shows how widely the values of that random variable are likely to be. The larger the variance, the more scattered the observations about its mean value. In other words, variance shows the concentration of the random variable's distribution with regard to its mean value. Variance of a continuous random variable X, denoted by σ^2 or $V(x)$, is defined by

$$\sigma^2 = V(x) := \int_{-\infty}^{\infty} (x - \mathbb{E}(x))^2 f(x) dx \tag{2.8}$$

The square root of the variance of a random variable is known as standard deviation (σ).

If the random variable X is discrete, then its variance $Var[X]$ is defined as

$$\begin{aligned} Var[X] &= \mathbb{E}[(X - \mu)^2] \\ &= \mathbb{E}[X^2 - 2\mu X + \mu^2] \\ &= \mathbb{E}[X^2] - 2\mu\mathbb{E}[X] + \mu^2 \\ &= \mathbb{E}[X^2] - \mu^2 \\ &= \mathbb{E}[X^2] - \mathbb{E}[X]^2. \end{aligned} \tag{2.9}$$

The variance and standard deviation of a random variable are always non-negative.

Covariance, Correlation and Correlation Coefficient

The *covariance* of two random variables X and Y shows how much X and Y change together. If these two random variables change values in the same direction, i.e. when X takes greater (or lesser) values the random variable Y takes greater (lesser) values, too, then the covariance of X and Y is positive. When the opposite scenario occurs, the covariance would be negative. The sign of the covariance shows the tendency in the linear relationship between X and Y.

The covariance $Cov(X, Y)$ of X and Y is defined as

$$\begin{aligned} Cov(X, Y) &= \mathbb{E}[(X - \mathbb{E}[X])(Y - \mathbb{E}[Y])] \\ &= \mathbb{E}[XY - X\mathbb{E}[Y] - \mathbb{E}[X]Y + \mathbb{E}[X]\mathbb{E}[Y]] \\ &= \mathbb{E}[XY] - \mathbb{E}[X]\mathbb{E}[Y] - \mathbb{E}[X]\mathbb{E}[Y] + \mathbb{E}[X]\mathbb{E}[Y] \\ &= \mathbb{E}[XY] - \mathbb{E}[X]\mathbb{E}[Y]. \end{aligned} \tag{2.10}$$

In statistics, *correlation* shows how strongly (or, weakly) two or more variables are related to each other. Linear correlation (also known as Pearson's correlation) determines the extent to which two variables are linearly related (or proportional) to each other.

Correlation coefficient denoted by ρ is a number. It indicates the degree to which two variables X and Y are linearly related. Let the mean and the standard deviation of X (Y) be denoted by μ_X (μ_Y) and σ_X (σ_Y), respectively. The correlation coefficient of X and Y, $\rho_{X,Y}$, can be expressed as

$$\begin{aligned}\rho_{X,Y} &= \frac{Cov(X,Y)}{\sigma_X \sigma_Y} \\ &= \frac{\mathbb{E}[(X-\mu_X)(Y-\mu_Y)]}{\sigma_X \sigma_Y} \\ &= \frac{\mathbb{E}[(X-\mathbb{E}(X))(Y-\mathbb{E}(Y))]}{\sqrt{\mathbb{E}(X^2)-\mathbb{E}^2(X)}\sqrt{\mathbb{E}(Y^2)-\mathbb{E}^2(Y)}} \\ &= \frac{\mathbb{E}(XY)-\mathbb{E}(X)\mathbb{E}(Y)}{\sqrt{\mathbb{E}(X^2)-\mathbb{E}^2(X)}\sqrt{\mathbb{E}(Y^2)-\mathbb{E}^2(Y)}}.\end{aligned} \tag{2.11}$$

If the two variables are independent then $\rho_{X,Y} = 0$ and if one variable (say Y) is a linear function of the other variable (X), then $\rho_{X,Y}$ is either 1 or -1. What should be noted is that the correlation (or correlation coefficient) measures linear relationship only. It is possible that there exists a strong nonlinear relationship between two variables while $\rho_{X,Y} = 0$. In the case of nonlinear relationship between two random variables, the correlation coefficient does not show the influence of one variable on the other.

Autocorrelation

A random variable X could be a process of the time. At each time instant, $X(t)$ (or, X_t) itself is a random variable. Such $X(\cdot)$ can be called a *random process*. More will be described shortly below. The term "autocorrelation" is the correlation of a signal process with itself (Phillips et al. 1995). It indicates the similarity between observations as a function of the time separation. In other words, autocorrelation of a random process describes the correlation between the values of the same process at different points in time. It is also sometimes called as *lagged* or *serial* correlation. Therefore, if X is a process with the mean μ and the standard deviation σ, both are functions of the time, then the autocorrelation of X, $\gamma(t,m)$ with regard to two different time instants t and m can be defined as

$$\gamma(t,m) := \frac{\mathbb{E}[(X_t-\mu_t)(X_m-\mu_m)]}{\sigma_t \sigma_m} \tag{2.12}$$

If the variance of this random process over the time is non-zero finite, then $\gamma(t, m)$ is well defined and takes value between 1 and −1. If X_t is a second-order stationary process (sometimes called *wide-sense stationary* (WSS) process), i.e. the mean and variance are time-independent, then its autocorrelation only depends on the time lapse between the pair of time instants. Therefore, if let $\tau := t - m$, (2.12) can be re-written as

$$\gamma(\tau) = \frac{\mathbb{E}[(X_t - \mu)(X_{t+\tau} - \mu)]}{\sigma^2} \tag{2.13}$$

Autocorrelation helps in finding repeating patterns, such as periodic changes or fundamental frequency of a signal which cannot be determined due to some unwanted factors like noise. A positive autocorrelation value indicates some sort of tendency for a system to remain in the same status from one time instant to another. For example, the likelihood of tomorrow being rainy is higher if today is raining than if today is dry (Chatfield 2016).

Normal Distribution

The "normal distribution" (also known as Gaussian distribution) is a very important class of statistical distribution. This type of distribution is symmetric about the mean value of the random variable and has a bell-shaped density curve with a single peak. Any normal distribution can be fully specified by two quantities: the mean μ (or m), where the peak of the density lies at, and the standard deviation σ, which indicates the spreading distribution. In other words, if a real-valued random variable X is normally distributed with mean m and variance σ^2, where $\sigma \geq 0$, then it can be written as

$$X \sim \aleph(m, \sigma^2) \tag{2.14}$$

i.e. the random variable X is normally distributed with mean m and variance σ^2. The normal probability density function of a scalar value can be shown analytically to be

$$f(x) = \frac{1}{\sigma\sqrt{2\pi}} e^{[-\frac{(x-m)^2}{2\sigma^2}]} \tag{2.15}$$

The integral of the above function (which corresponds to the area under the curve) is unity. Sketch of a typical normal distribution is shown in Fig. 2.2 for $\sigma = 4$ and $m = 13$.

The probability that a normal distributed function lies outside $\pm 2\sigma$ is approximately 0.05. Standard normal distribution is defined with mean $m = 0$ and standard deviation $\sigma = 1$. Therefore, from Eq. (2.15), a standard normal distribution can be written as

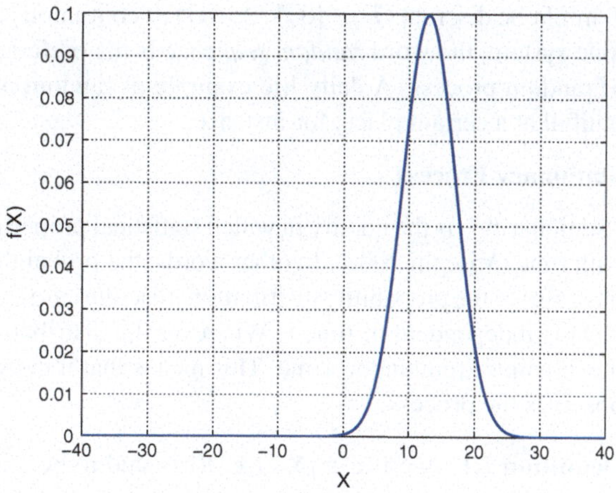

Fig. 2.2 Normal probability density function

$$f(x) = \frac{1}{\sqrt{2\pi}} e^{[-\frac{x^2}{2}]} \tag{2.16}$$

The normal distribution (2.15) can be extended to the cases of two and more random variables. For the case of two random variables, the bivariate normal distribution is as below:

$$f(x_1, x_2) = \frac{1}{\sigma_1\sigma_2\sqrt{1-\rho^2}} e^{\left[-\frac{\frac{x_1^2}{\sigma_1^2} - 2\rho\frac{x_1 x_2}{\sigma_1\sigma_2} + \frac{x_2^2}{\sigma_2^2}}{2(1-\rho^2)}\right]} \tag{2.17}$$

where ρ is the correlation function between the two random variables x_1 and x_2. Similarly, for the case of n random variables, the multivariate normal distribution function is

$$f(x_1, x_2, \ldots, x_n) = \frac{1}{(2\pi)^{n/2} \| P \|^{1/2}} e^{[-\frac{(\mathbf{x}-\mathbf{m})^T P^{-1} (\mathbf{x}-\mathbf{m})}{2\sigma^2}]} \tag{2.18}$$

where the vector $\mathbf{x}^T := [x_1, x_2, \ldots, x_n]$; and, the vector $\mathbf{m} := \mathbb{E}[\mathbf{x}]$ and the matrix $P := \mathbb{E}[(\mathbf{x}-\mathbf{m})(\mathbf{x}-\mathbf{m})^T]$ are the mean and covariance of the vector $\mathbf{x}$, respectively.

Random Process

A random or stochastic process is a process whose behaviour is not completely predictable. A random process can, however, be characterised by its statistical properties. Due to randomness of random variables, the average values from a collection of signals rather than from one individual signal, are usually investigated. Therefore, a random process $\{X(t), t \in T\}$ is a family of random variables where the index set

T might be discrete ($T = \{0, 1, 2, \ldots\}$) or continuous ($T = (0, \infty)$). For a discrete-time system, the word *random sequence* is the preferred terminology to use instead of random process. A daily life example of random processes could be the hourly rainfall at a certain place, for instance.

Stationary Process

"Stationarity" is the quality in which statistical properties of a process do not change with time (Vaseghi 2008). In other words, the probability density function (continuous case) or the probability distribution (discrete case) of a stationary random variable $X(t)$ is independent of time t. Wherever the distribution is seen for some segment, the dynamics remain the same. This means that it does not matter when in time one observes the process.

Definition 2.1 A process $\{X_t, t \in \Re\}$ is said to be a *weak-sense stationary* (WSS) process if

1. $\mathbb{E}[X_t^2] < \infty, \forall t \in \Re$,
2. $\mathbb{E}[X_t] = \mu_t = \mu$, and
3. $Cov(X_t, X_m) = \mathbb{E}[(X_t - \mu_t)(X_m - \mu_m)] = \sigma^2(t, m) = \sigma^2(t - m), \forall t, m \in \Re$.

In other words one can say that a weak-sense stationary process must have finite variance, constant mean (first moment) and the second moment should only depend on $(t - m)$ and not on t or m (Vaidyanathan 2007).

Definition 2.2 A process $\{X_t, t \in \Re\}$ is *strictly stationary* process if all its finite (cumulative) distribution function is independent of the time shift, i.e. for any finite number n and $\forall t_1, t_2, \ldots, t_n \in \Re, \forall h \in \Re$,

$$\begin{aligned} p(X_{t_1} \leq x_1, X_{t_2} \leq x_2, \ldots, X_{t_n} \leq x_n) &= F(x_{t_,} x_{t_2}, \ldots, x_{t_n}) \\ &= F(x_{t_1+h}, x_{t_2+h}, \ldots, x_{t_n+h}) \\ &= p(X_{t_1+h} \leq x_1, X_{t_2+h} \leq x_2, \ldots, X_{t_n+h} \leq x_n) \end{aligned} \quad (2.19)$$

Most statistical prediction methods are based on the assumption that the time series can be considered approximately stationary.

White Noise

White noise is a random signal (or process) in which the power spectrum is distributed uniformly over all the frequency components in the full infinite range. This is purely a theoretical concept because if a signal has the same power at all components frequency, this is equivalent to a signal whose total power is infinite and therefore it is impossible to generate such a signal in the real world. However in practice, for a finite frequency range, a "white" signal with a flat spectrum can be assumed.

Mathematically, a random signal (or, a vector signal) $\mathbf{v}$ is a white signal if and only if it possesses the following two properties:

$$m_{\mathbf{v}} := \mathbb{E}[\mathbf{v}] = 0$$
$$R_{\mathbf{v}} := \mathbb{E}[\mathbf{v}\mathbf{v}^T] = \sigma^2 I \tag{2.20}$$

A continuous white noise process $\mathbf{w}$ holds the same properties as

$$m_{\mathbf{w}}(t) := \mathbb{E}[\mathbf{w}(t)] = 0$$
$$Q_{\mathbf{w}(t_1,t_2)} := \mathbb{E}[\mathbf{w}(t_1)\mathbf{w}(t_2)^T] = \frac{1}{2}S\delta(t_1 - t_2) \tag{2.21}$$

i.e. it has zero mean everywhere and infinite power at zero time shift.

2.3 Desired Properties of State Estimators

Obviously, the purpose of using a state estimator is to obtain an estimate of the system states. That is to produce an $\hat{x}(t)$ such that it reflects the true value (information) of the actual state $x(t)$. This property would usually be referred as "convergence": the estimate $\hat{x}(t)$ needs to converge to the actual state $x(t)$ or in almost all state estimation cases involving stochastic signals (noises), as in this book, the expectation of the magnitude of the difference between $x(t)$ and $\hat{x}(t)$ should be zero. For a vector state, it turns out to be $\mathbb{E}\left[\|x(t) - \hat{x}(t)\|^2\right] = 0$, in terms of so-called the mean square error (MSE). However, in the context of filtering, this property is more often called ***stability***. It indicates that the convergence property comes from the asymptotic stability of the dynamic system governing the estimation error $e(t) = x(t) - \hat{x}(t)$. Such use of this terminology could trace back to the classical state observer like the Luenberger observer discussed in the next section. With the term of stability, the concept of convergence rate can also be linked to various stability definitions, such as exponential stability.

Many state estimation algorithms consist of, like most approaches discussed in this book, two phases in operation: predict and update. In the prediction phase, the states and covariance functions are predicted based on the (known) system dynamics, *i.e.* the state evolution functions in a state-space model as well as on the estimated states, the covariances at the previous time instant, and the system input information at the current time. In the updating phase, with the newly obtained measurements, the state estimates and covariance are updated. In reality, system dynamics may not be known exactly nor be truly incorporated in the system model. Also, the statistic properties of the noises involved in the system running such as their covariances may not be known exactly. These would seriously influence the effectiveness of state estimation algorithms. Thus, how well a state estimation algorithm can deal with system dynamics and noises variations is important and worth considering. This property is usually referred to as ***sensitivity***. It is part of a broader property called ***robustness***. The robustness consideration may also include the scenario of measurement missing, for instance, that will be discussed later in this book.

Applicability and computational efficiency are among other features requested for a state estimation algorithm. For example, the case of non-Gaussian noises needs to be considered in a practical application; and, approaches such as the square root form of algorithms may be preferable in solving a real-world problem for numerical stability. Interested readers will find a great amount of discussions on these features in the literature (Allison 2002; Blind et al. 2009; Khan and Gu 2009; Khan et al. 2010; Khan 2011).

2.4 Least Square Estimator and Luenberger State Observer

Often, when data (or, measurements) are corrupted by some unwanted (noise) signal, the information contained in the data will have to be obtained by the means of filtering (estimation). In the context of control systems, to get the information of the state is the process of state estimation. In a wider sense, this is also a process of determination of system "parameters" from available data. Hence, it may also be called "parameter estimation". A well-known optimal parameter estimator is next introduced in this section which bears the minimum prediction (or estimation) error, and is called "least square estimator" (LSE). Before the introduction of LSE, linear control system models will be presented below first.

Linear Control System Models

It is a fact that almost all control systems in the real world would have certain nonlinearities, in the form of inherent nonlinear dynamics, input saturation or others. However, it is also true that many nonlinear systems can be approximated by linear system models, though locally in most cases. Obviously it is much more convenient and easier to analyse a linear system and to conduct control scheme design for a linear system. Analysis tools and synthesis approaches for linear control systems are also much more mature. Of course, one has to bear in mind in the process of analysis and design around a linear model the nonlinear nature of the underlying control system, if the actual system is nonlinear which is the usual case.

For linear, finite-dimensional, time-invariant, continuous-time control systems, a state-space model is of the form

$$\dot{\mathbf{x}}(t) = A\mathbf{x}(t) + B\mathbf{u}(t)$$
$$\mathbf{y}(t) = C\mathbf{x}(t) + D\mathbf{u}(t)$$

where A, B, C and D are constant matrices with appropriate dimensions.

Equation (2.22) is called the state equation and (2.22) the output equation. The state equation is a set of linear, ordinary differential equations with constant coefficients; the output $\mathbf{y}(t)$ is a *memoryless* function of $\mathbf{x}(t)$ and $\mathbf{u}(t)$. In many systems, $D = 0$.

Least Square Estimation

Linear least square estimation is an estimation method for unknown parameters in a linear regression model with the aim of minimising the sum of the square of differences between the true parameters and the estimated parameters. Consider an unknown parameter vector x that is related to the data or observation vector z in the following equation:

$$\mathbf{z} = H\mathbf{x} + \mathbf{v} \tag{2.22}$$

This could be the case of the output equation (2.22) with $D = 0$ of the linear, time-invariant system model introduced earlier. One may assume that the measuring process of the system output is contaminated by a noise $\mathbf{v}$, that is a common feature in the real operation environment. The measured signal (also called the measurement or observation) $\mathbf{z}$ is then given by $\mathbf{z} = \mathbf{y} + \mathbf{v} = C\mathbf{x} + \mathbf{v}$.

The aim is to have an estimation $\hat{\mathbf{x}}$ of the variable $\mathbf{x}$ that minimises some "measure" of the estimation error $\epsilon := \mathbf{z} - H\hat{\mathbf{x}}$. The measure of the estimation error can be characterised by the squared Euclidean vector norm, i.e. to minimise the scalar cost function J, where

$$J = (z - H\hat{\mathbf{x}})^T (z - H\hat{\mathbf{x}}) \tag{2.23}$$

Candidates for $\hat{\mathbf{x}}$ to minimise this scalar cost function can be obtained by satisfying the following equation:

$$\frac{\partial J}{\partial \hat{\mathbf{x}}} = 0 \tag{2.24}$$

By substituting (2.23) in (2.24), straightforward manipulations lead to

$$H^T H\hat{\mathbf{x}} = H^T \hat{\mathbf{x}} \tag{2.25}$$

Equation (2.25) is sometimes called the *normal equation* for the least square problem (Grewal and Andrews 2001). It can be solved by, provided that the matrix $H^T H$ is nonsingular,

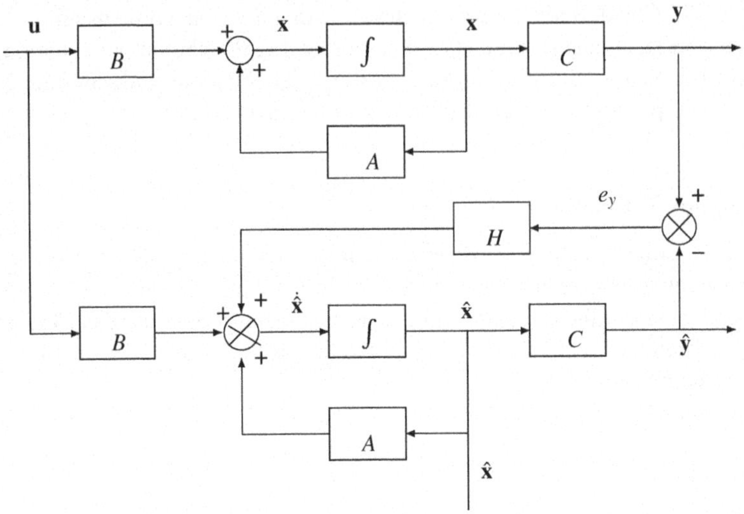

Fig. 2.3 Luenberger (full-order) state observer

$$\hat{\mathbf{x}} = (H^T H)^{-1} H^T z \tag{2.26}$$

Luenberger State Observer

Probably the most well-known state estimator for linear, time-invariant (LTI) systems is the Luenberger state observer. Consider again the LTI system model (2.22). The target is to get a "good" estimation of the most likely internal information, the state **x**. Directly available information is the input **u** and output **y**. The system matrices $(A,\ B,\ C)$ are supposed to be known, with D being assumed to be zero. It will not be possible to reconstruct the state **x** based on the system matrices and the input, by either hardware or computer programs, because of the initial states being unknown. One can, however, build up a Luenberger state observer to generate an estimation of the system state, as in the following.

The basic idea of the state observer is to construct the state-space model of the control system, usually using computer algorithms, and drive this system model (observer) with the same input signal applied to the real system. Due to the unknown initial states of the real system, the states of the observer will not coincide with the actual states. The output of the observer will then be compared with that of the actual system which is always available. The difference will be fed back, through a gain matrix, to the observer. By appropriately choosing the gain matrix, the difference between the estimated states and the actual states will converge to zero, and therefore the estimated states can be taken as a good approximation of the true states.

The above can be seen clearly in the following block diagram. The bottom part is the observer, of which the state is denoted as $\hat{\mathbf{x}}$ and the output $\hat{\mathbf{y}}$, and the gain matrix H is to be assigned (Fig. 2.3).

It is a direct exercise to obtain the state-space model of the observer as the following.

$$\frac{d\hat{\mathbf{x}}}{dt} = A\hat{\mathbf{x}} + B\mathbf{u} + H(\mathbf{y} - \hat{\mathbf{y}})$$
$$\hat{\mathbf{y}} = C\hat{\mathbf{x}}$$

Define the state difference as $\mathbf{e} = \mathbf{x} - \hat{\mathbf{x}}$. The difference $\mathbf{e}$ is governed by the following dynamic equation

$$\begin{aligned}\dot{\mathbf{e}} &= \frac{d}{dt}(\mathbf{x} - \hat{\mathbf{x}}) \\ &= A(\mathbf{x} - \hat{\mathbf{x}}) - HC(\mathbf{x} - \hat{\mathbf{x}}) \\ &= (A - HC)\mathbf{e}\end{aligned}$$

The above reveals clearly that if the (output) gain matrix H is chosen such that the matrix $(A - HC)$ is stable, i.e. all its eigenvalues are in the open left-half-plane, then the state difference $\mathbf{e}$ will converge to zero for any initial $\mathbf{e(0)}$, as $t \to \infty$, which of course indicates $\hat{\mathbf{x}} \to \mathbf{x}$. Therefore, a key point in the construction of a state observer is that the chosen H should make $(A - HC)$ stable. A question remains: what kind of system would assure the existence of such H? The next theorem answers it and its proof can be readily found in most control course textbooks that cover the state-space approach.

Theorem 2.1 *The eigenvalues of* $(A - HC)$ *can be arbitrarily assigned* if and only if *the matrix pair* $(A,\ C)$ *is* completely observable.

Luenberger state observer introduced above is for the continuous-time systems. Similar development exists for discrete-time systems, too.

2.5 Luenberger State Observer for a DC Motor

Consider a second-order DC motor (linearised) model with the current and angular velocity as its states, $x = [i \quad \omega]^T$ and the current as the output (Mehta and Chiasson 1998),

$$\dot{i} = -\frac{Ri + K_T i\omega - V}{L} \tag{2.27}$$

$$\dot{\omega} = K_T L i^2 - D_m \omega \tag{2.28}$$

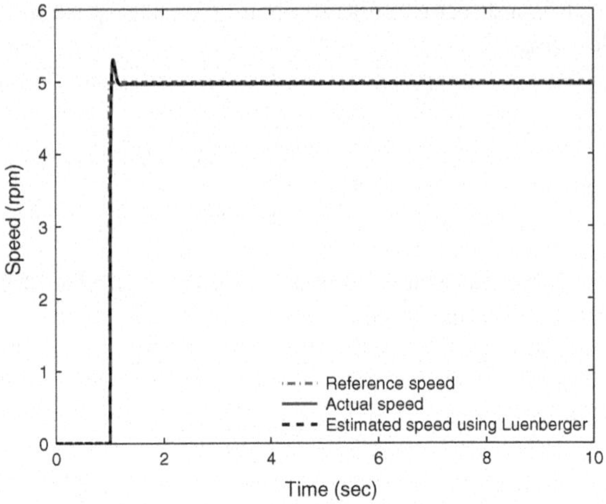

Fig. 2.4 States of Luneberger observer for dc motor

where, $R = 7.2\Omega$, $L = 0.5H$, $D_m = 0.0004$, $J = 0.00071$ and $K_T = 0.01$. By linearising the above model, the state-space matrices can be obtained as:

$$A = \begin{bmatrix} -\frac{R+K_T\omega_0}{L} & -\frac{K_T}{Li_0} \\ \frac{2K_Ti_0}{J} & -\frac{D_m}{J} \end{bmatrix}$$
$$B = \begin{bmatrix} \frac{1}{L} \\ 0 \end{bmatrix}$$
$$C = \begin{bmatrix} 1 \; 0 \end{bmatrix}$$
$$D = 0$$

The operating points are $i_0 = 0.1A$ and $\omega_0 = 1rad/s$. The reference velocity and the output velocity are compared and the error is sent to a PID controller with $K_p = 8.8$, $K_i = 267.9$ and $K_d = 0.03$. The output of the designed PID controller is taken as the input to the DC motor. However, it is assumed that only the current of the DC motor is measured and the velocity is estimated using a Luneberger observer. The Luenberger observer gain, H, is designed such that the eigenvalues of $(A - HC)$ are placed at -1 and -20, respectively. The corresponding Luenberger gain for the DC motor is obtained as $H = [6 \quad -4240.39]^T$ and is then used in (2.27) to estimate the angular velocity. Instead of the actual velocity, the estimated velocity from Luenberger observer is compared with the reference velocity to generate the error which is then given to the PID controller. Figure 2.4 shows the reference velocity, actual velocity and the velocity obtained by using the Luenberger observer. The actual and estimated velocity are very close to each other; the error between them is shown in Fig. 2.5.

The above simulations shows how well the estimated signal is close to the actual signal. It also shows of course the successful design of the PID controller which makes

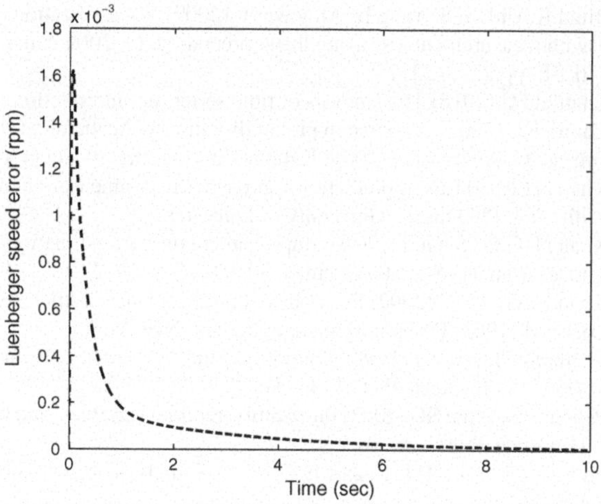

Fig. 2.5 Luneberger observer estimation error for dc motor

the output of the DC motor closely follow the reference signal. In this example, the nonlinearity of the DC motor is relatively "mild". The linearised model is thus "close" to the nonlinear model at the chosen operating point (also called the equilibrium point) and the Luenberger observer therefore works well in such cases. Readers may notice as well that no noises are supposed to affect the running of the motor. Interested readers may play with the observer gain and/or the PID controller parameter values to see the efficiency of the observer and of the controller designated, too.

2.6 Summary

This is a preliminary chapter. It introduced some mathematical components which are necessary for the development of state estimation approaches to be discussed in the rest of the book. It also introduced the basic ideas of state estimators, their usage and their importance in control systems analysis and control schemes design. This chapter also introduced the classical state estimator, *Luenberger state observer*, with a DC motor example. Luenberger state observer is widely used in re-construction of states for a linear and time-invariant control system which is deterministic and is supposed to be free from noise contamination. Luenberger state observer and the application example offered a taste for the simplest case of system states estimation.

References

Allison PD (2002) Missing data: quantitative applications in the social sciences. Br J Math Stat Psychol 55(1):193–196

Blind R, Uhlich S, Yang B, Allgower F (2009) Robustification and optimization of a kalman filter with measurement loss using linear precoding. In: 2009 American Control Conference, ACC'09, IEEE, pp 2222–2227

Chatfield C (2016) The analysis of time series: an introduction. CRC Press, Boca Raton

Chung KL (2001) A course in probability theory. Academic, Cambridge

Grewal M, Andrews A (2001) Kalman filtering: theory and practice using matlab

Khan N (2012) Linear prediction approaches to compensation of missing measurements in kalman filtering. PhD thesis, University of Leicester

Khan N, Fekri S, Gu D (2010) Improvement on state estimation for discrete-time LTI systems with measurement loss. Measurement 43(10):1609–1622

Khan N, Gu D-W (2009) Properties of a robust kalman filter. IFAC Proc 42(19):465–470

Loeve M (1963) Probability theory, 3rd edn. New York

Mehta S, Chiasson J (1998) Nonlinear control of a series DC motor: theory and experiment. IEEE Trans Ind Electron 45(1):134–141

Papoulis A, Pillai SU (2002) Probability, random variables, and stochastic processes. Tata McGraw-Hill Education, New York

Phillips CL, Parr JM, Riskin EA (1995) Signals, systems, and transforms. Prentice Hall, Upper Saddle River

Vaseghi SV (2008) Advanced digital signal processing and noise reduction. Wiley, New Jersey

Vaidyanathan P (2007) The theory of linear prediction. Synth Lect Signal Process 2 (1):1–184

Wiener N (1949) Extrapolation, interpolation, and smoothing of stationary time series, vol 7. MIT press, Cambridge

Chapter 3
Kalman Filter and Linear State Estimations

3.1 Introductory Remarks

As discussed in the earlier chapters, the system state contains vital information for control system analysis and design, but it is not always directly measurable. Control engineers will have to obtain the information of system states based on accessible information of the known system dynamics, inputs and outputs (measurements). What makes this task more complicated is that industrial control systems are always operating in an environment with uncertainties. Uncertainties include dynamic perturbations and unwanted, as well as unknown, external signals such as noises. Noises affect the operation of sensors and transducers, and thus generate inaccurate measurements. That would give wrong information about the states of the control system and even possibly lead to failure of a control scheme in a feedback system structure. To retrieve the actual, true information from noise-contaminated signals is the target of the filtering (estimation) process. The study on state estimation has been an active research subject for several decades and always plays a key role in practical applications. Real-life applications of the state estimation include state and parameter estimation in chemical process plants, electrical machinery, data assimilation, econometrics, fault detection and isolation, control system design, etc., where either sensors are noisy or it is difficult to measure the states due to various physical settings. Linear state estimation deals with the state estimation of linear systems and linear measurement models. Although, all the practical systems have nonlinearities involved as discussed earlier and nonlinear system state estimation is the major theme of this book, it is worth to explore some linear state estimation methods as a basis for nonlinear estimation approaches. In this chapter, the epic development, the Kalman filter (KF), will be introduced followed by the information filter and the $\mathcal{H}_\infty$ filters. Similar to the previous chapter, and the following chapters as well, applications of these filters will be demonstrated at a later part of the chapter. One may see Chandra (2013) for more details.

K. P. B. Chandra and D.-W. Gu, *Nonlinear Filtering*,
https://doi.org/10.1007/978-3-030-01797-2_3

3.2 The Discrete-Time Kalman Filter

In early 1940s Kolmogorov and Wiener discussed the problem of linear least square estimation for a stochastic process (Kolmogorov 1962; Wiener 1949; Kailath 1968; Kailath et al. 2000). Kolmogorov formulated the filter design problem using statistical and frequency domain characteristics. However their results are based on stationary processes with infinite or semi-infinite observation intervals. R.E. Kalman presented the design of now well-known *Kalman filter* (KF) in Kalman et al. (1960). Kalman filtering approach is an online, recursive method adopted to estimate the state of a system with noise-contaminated observations. Kalman filter is to be introduced in this section next.

The signals involved in an industrial control system (such as input, output and state) are most likely continuous-time signals. However, due to the wide use of digital computers nowadays in control engineering and due to the recursive (iterative) nature of most estimation schemes, control systems, and consequently the state estimators, considered in this book will be modelled and discussed in the discrete-time context.

3.2.1 *Process and Measurement Models*

Consider the discrete linear process and measurement models as

$$\mathbf{x}_k = \mathbf{F}_{k-1}\mathbf{x}_{k-1} + \mathbf{G}_{k-1}\mathbf{u}_{k-1} + \mathbf{w}_{k-1}, \tag{3.1}$$

$$\mathbf{z}_k = \mathbf{H}_k\mathbf{x}_k + \mathbf{v}_k, \tag{3.2}$$

where $k \geq 1$ is the time index, $\mathbf{x}_k$ is the state vector, $\mathbf{u}_k$ the control input, $\mathbf{z}_k$ the measurement (output), $\mathbf{w}_{k-1}$ and $\mathbf{v}_k$ are the process and measurement noises, respectively.

Noises are *random* signals. A control system with stochastic components can be called a stochastic system. Strictly speaking though, the dynamics of a stochastic system should be a stochastic processes. The control systems considered in this book are mainly those with deterministic dynamics but influenced by stochastic signals namely noises.

The process and measurement noises are assumed to be Gaussian-distributed[1] random variables, which have zero means and covariances of $\mathbf{Q}_{k-1}$ and $\mathbf{R}_k$, respectively,

---

[1]The Kalman's original derivation did not use the Bayes' rule and does not require the exploitation of any specific error distribution information. The Kalman filter is the minimum variance estimator if the noise is Gaussian, and it is the ***linear*** minimum variance estimator for linear systems with non-Gaussian noises (Kalman et al. 1960; Simon 2006; Julier and Uhlmann 2004; Van Der Merwe 2004).

$$\mathbf{w}_{k-1} = \mathcal{N}(\mathbf{0}, \mathbf{Q}_{k-1})$$
$$\mathbf{v}_k = \mathcal{N}(\mathbf{0}, \mathbf{R}_k)$$

where $\mathcal{N}$ represents the Gaussian or normal probability distribution.

The process and measurement noises at any time are assumed to be independent of the system state

$$\mathbb{E}\,[\mathbf{w}_i \mathbf{x}_j^T] = \mathbf{0}, \quad \mathbb{E}\,[\mathbf{w}_i \mathbf{w}_j^T] = \mathbf{Q}_i \delta_{ij} \quad \forall \;\; i, j$$

and

$$\mathbb{E}\,[\mathbf{v}_i \mathbf{x}_j^T] = \mathbf{0}, \quad \mathbb{E}\,[\mathbf{v}_i \mathbf{v}_j^T] = \mathbf{R}_i \delta_{ij} \quad \forall \;\; i, j$$

where δ_{ij} is the Dirac function and $\mathbb{E}\,[\cdot]$ is the expectation operator.

3.2.2 Derivation of the Kalman Filter

The derivation in this section closely follows the one given in Csorba (1998) and Durrant-Whyte et al. (2001). The Kalman filter gives an estimate, $\hat{\mathbf{x}}_{i|j}$, that minimises the mean-squared estimation error conditioned on the measurements sequence, $\mathbf{Z}_j = [\mathbf{z}_1, \mathbf{z}_2, \ldots, \mathbf{z}_j]$. The estimated state is the expected value of the state conditioned on the measurements sequence and is given by

$$\hat{\mathbf{x}}_{i|j} = \mathbb{E}[\mathbf{x}_i | \mathbf{Z}_j]. \tag{3.3}$$

The error between the actual and estimated states is defined by

$$\mathbf{x}_{e,i|j} = \mathbf{x}_i - \hat{\mathbf{x}}_{i|j} \tag{3.4}$$

and the covariance of $\hat{\mathbf{x}}_{i|j}$

$$\mathbf{P}_{i|j} = \mathbb{E}[\mathbf{x}_{e,i|j} \mathbf{x}_{e,i|j}^T | \mathbf{Z}_j]. \tag{3.5}$$

The Kalman filter is a recursive algorithm consisting of two stages: the time update (also known as a priori or prediction) stage, and the measurement update (a posteriori estimate or filtering update) stage. In the prediction stage, the state and the covariance at the kth instant are predicted based on the information available at the $(k-1)$th instant. Once the (new) measurement is obtained at the kth instant, the predicted state and covariance are then "modified" to yield the updated state estimate and updated covariance at the kth instant. The whole process repeats itself recursively.

The predicted state can be obtained by taking the expectation of the state model conditioned on measurements up to the $(k-1)$th instant and is given by

$$\begin{aligned}\hat{\mathbf{x}}_{k|k-1} &= \mathbb{E}[\mathbf{x}_k|\mathbf{Z}_{k-1}]\\ &= \mathbb{E}[(\mathbf{F}_{k-1}\mathbf{x}_{k-1} + \mathbf{G}_{k-1}\mathbf{u}_{k-1} + \mathbf{w}_{k-1})|\mathbf{Z}_{k-1}]\\ &= \mathbf{F}_{k-1}\mathbb{E}[\mathbf{x}_{k-1}|\mathbf{Z}_{k-1}] + \mathbf{G}_{k-1}\mathbf{u}_{k-1} + \mathbb{E}[\mathbf{w}_{k-1}|\mathbf{Z}_{k-1}]\\ &= \mathbf{F}_{k-1}\hat{\mathbf{x}}_{k-1|k-1} + \mathbf{G}_{k-1}\mathbf{u}_{k-1}.\end{aligned} \tag{3.6}$$

Similarly, the predicted covariance at the kth instant based on the $(k-1)$th can be found as

$$\begin{aligned}\mathbf{P}_{k|k-1} &= \mathbb{E}[(\mathbf{x}_{e,k|k-1}\mathbf{x}_{e,k|k-1}^T)|\mathbf{Z}_{k-1}]\\ &= \mathbb{E}[(\mathbf{x}_k - \widehat{\mathbf{x}}_{k|k-1})(\mathbf{x}_k - \widehat{\mathbf{x}}_{k|k-1})^T|\mathbf{Z}_{k-1}]\\ &= \mathbb{E}[(\mathbf{F}_{k-1}\mathbf{x}_{k-1} - \mathbf{F}_{k-1}\widehat{\mathbf{x}}_{k-1|k-1} + \mathbf{w}_{k-1})(\mathbf{F}_{k-1}\mathbf{x}_{k-1} - \mathbf{F}_{k-1}\widehat{\mathbf{x}}_{k-1|k-1} + \mathbf{w}_{k-1})^T|\mathbf{Z}_{k-1}]\\ &= \mathbf{F}_{k-1}\mathbb{E}[(\mathbf{x}_{k-1} - \widehat{\mathbf{x}}_{k-1|k-1})(\mathbf{x}_{k-1} - \widehat{\mathbf{x}}_{k-1|k-1})^T|\mathbf{Z}_{k-1}]\mathbf{F}_{k-1}^T + \mathbb{E}[(\mathbf{w}_{k-1}\mathbf{w}_{k-1}^T)|\mathbf{Z}_{k-1}]\\ &= \mathbf{F}_{k-1}\mathbf{P}_{k-1|k-1}\mathbf{F}_{k-1}^T + \mathbf{Q}_{k-1}.\end{aligned} \tag{3.7}$$

The predicted measurement and the *innovation* vector, which is the difference between actual measurement and the predicted measurement, can be obtained as

$$\begin{aligned}\widehat{\mathbf{z}}_{k|k-1} &= \mathbb{E}[\mathbf{z}_k|\mathbf{Z}_{k-1}]\\ &= \mathbb{E}\big[\mathbf{H}_k\mathbf{x}_k + \mathbf{v}_k|\mathbf{Z}_{k-1}\big]\\ &= \mathbf{H}_k\mathbb{E}[\mathbf{x}_k|\mathbf{Z}_{k-1}] + \mathbb{E}[\mathbf{v}_k|\mathbf{Z}_{k-1}]\\ &= \mathbf{H}_k\widehat{\mathbf{x}}_{k|k-1}\end{aligned} \tag{3.8}$$

and

$$\begin{aligned}\nu_k &= \mathbf{z}_k - \widehat{\mathbf{z}}_{k|k-1}\\ &= \mathbf{z}_k - \mathbf{H}_k\widehat{\mathbf{x}}_{k|k-1}.\end{aligned} \tag{3.9}$$

After obtaining the predicted state and covariance, the next task is to obtain the updated state and covariance for the recursive process. The updated state vector can be calculated from the predicted state and the innovation vector and is given as

$$\widehat{\mathbf{x}}_{k|k} = \widehat{\mathbf{x}}_{k|k-1} + \mathbf{K}_k\nu_k \tag{3.10}$$

where, $\mathbf{K}_k$ is the Kalman gain, which dictates the influence of the innovation on the updated state vector. The error between the actual and updated states is given by

$$\begin{aligned}
\mathbf{x}_{e,k|k} &= \mathbf{x}_k - \widehat{\mathbf{x}}_{k|k} \\
&= \mathbf{x}_k - [\widehat{\mathbf{x}}_{k|k-1} + \mathbf{K}_k \nu_k] \\
&= \mathbf{x}_{e,k|k-1} - \mathbf{K}_k \nu_k \\
&= \mathbf{x}_{e,k|k-1} - \mathbf{K}_k(\mathbf{z}_k - \mathbf{H}_k \widehat{\mathbf{x}}_{k|k-1}) \\
&= \mathbf{x}_{e,k|k-1} - \mathbf{K}_k(\mathbf{H}_k \mathbf{x}_k + \mathbf{v}_k - \mathbf{H}_k \widehat{\mathbf{x}}_{k|k-1}) \\
&= \mathbf{x}_{e,k|k-1} - \mathbf{K}_k(\mathbf{H}_k \mathbf{x}_{e,k|k-1} + \mathbf{v}_k) \\
&= (\mathbf{I} - \mathbf{K}_k \mathbf{H}_k)\mathbf{x}_{e,k|k-1} - \mathbf{K}_k \mathbf{v}_k
\end{aligned} \tag{3.11}$$

where, $\mathbf{I}$ is the identity matrix of an appropriate size.
The covariance update can be written as

$$\begin{aligned}
\mathbf{P}_{k|k} &= \mathbb{E}[(\mathbf{x}_k - \widehat{\mathbf{x}}_{k|k})(\mathbf{x}_k - \widehat{\mathbf{x}}_{k|k})^T|\mathbf{Z}_k] \\
&= \mathbb{E}[\mathbf{x}_{e,k|k}\mathbf{x}_{e,k|k}^T|\mathbf{Z}_k] \\
&= \mathbb{E}[\{(\mathbf{I} - \mathbf{K}_k\mathbf{H}_k)\mathbf{x}_{e,k|k-1} - \mathbf{K}_k\mathbf{v}_k\}\{(\mathbf{I} - \mathbf{K}_k\mathbf{H}_k)\mathbf{x}_{e,k|k-1} - \mathbf{K}_k\mathbf{v}_k\}^T|\mathbf{Z}_k] \\
&= \mathbb{E}[(\mathbf{I} - \mathbf{K}_k\mathbf{H}_k)\mathbf{x}_{e,k|k-1}\mathbf{x}_{e,k|k-1}^T(\mathbf{I} - \mathbf{K}_k\mathbf{H}_k)^T - (\mathbf{I} - \mathbf{K}_k\mathbf{H}_k)\mathbf{x}_{e,k|k-1}\mathbf{v}_k^T\mathbf{K}_k^T \\
&\quad -\mathbf{K}_k\mathbf{v}_k\mathbf{x}_{e,k|k-1}^T(\mathbf{I} - \mathbf{K}_k\mathbf{H}_k)^T + \mathbf{K}_k\mathbf{v}_k\mathbf{v}_k^T\mathbf{K}_k^T|\mathbf{Z}_k] \\
&= (\mathbf{I} - \mathbf{K}_k\mathbf{H}_k)\mathbb{E}[\mathbf{x}_{e,k|k-1}\mathbf{x}_{e,k|k-1}^T](\mathbf{I} - \mathbf{K}_k\mathbf{H}_k)^T - (\mathbf{I} - \mathbf{K}_k\mathbf{H}_k)\mathbb{E}[\mathbf{x}_{e,k|k-1}\mathbf{v}_k^T|\mathbf{Z}_k]\mathbf{K}_k^T \\
&\quad -\mathbf{K}_k\mathbb{E}[\mathbf{v}_k\mathbf{x}_{e,k|k-1}^T|\mathbf{Z}_k](\mathbf{I} - \mathbf{K}_k\mathbf{H}_k)^T + \mathbf{K}_k\mathbb{E}[\mathbf{v}_k\mathbf{v}_k^T|\mathbf{Z}_k]\mathbf{K}_k^T \\
&= (\mathbf{I} - \mathbf{K}_k\mathbf{H}_k)\mathbf{P}_{k|k-1}(\mathbf{I} - \mathbf{K}_k\mathbf{H}_k)^T + \mathbf{K}_k\mathbf{R}_k\mathbf{K}_k^T \\
&= \mathbf{P}_{k|k-1} - \mathbf{K}_k\mathbf{H}_k\mathbf{P}_{k|k-1} - \mathbf{P}_{k|k-1}\mathbf{H}_k^T\mathbf{K}_k^T + \mathbf{K}_k\mathbf{R}_k\mathbf{K}_k^T.
\end{aligned} \tag{3.12}$$

By taking the trace[2] on both sides of Eq. (3.12) it yields

$$Tr(\mathbf{P}_{k|k}) = Tr(\mathbf{P}_{k|k-1}) - 2Tr(\mathbf{K}_k\mathbf{H}_k\mathbf{P}_{k|k-1}) + Tr(\mathbf{K}_k(\mathbf{H}_k\mathbf{P}_{k|k-1}\mathbf{H}_k^T)\mathbf{K}_k^T) + Tr(\mathbf{K}_k\mathbf{R}_k\mathbf{K}_k^T) \tag{3.13}$$

To evaluate the updated covariance, $\mathbf{P}_{k|k}$, the Kalman gain $\mathbf{K}_k$ is also required. The next task is to find the Kalman gain. The Kalman filter aims at minimising the mean-square estimation error. Taking the partial derivative[3] of Eq. (3.12) with respect to $\mathbf{K}_k$ gives

---

[2]

$$\begin{aligned}
(\mathbf{P}_{k|k-1}\mathbf{H}_k^T\mathbf{K}_k^T)^T &= \mathbf{K}_k\mathbf{H}_k\mathbf{P}_{k|k-1} \\
Tr(\mathbf{P}_{k|k-1}\mathbf{H}_k^T\mathbf{K}_k^T)^T &= Tr(\mathbf{K}_k\mathbf{H}_k\mathbf{P}_{k|k-1})
\end{aligned}$$

[3]For any matrix, M, and a symmetric matrix, N,

$$\frac{\partial Tr(MNM^T)}{\partial M} = 2MN.$$

$$\frac{\partial Tr(\mathbf{P}_{k|k})}{\partial \mathbf{K}_k} = -2\mathbf{P}_{k|k-1}\mathbf{H}_k^T + 2\mathbf{K}_k\mathbf{H}_k\mathbf{P}_{k|k-1}\mathbf{H}_k^T + 2\mathbf{K}_k\mathbf{R}_k. \tag{3.14}$$

Further, equating the right hand side of Eq. (3.14) to 0 yields

$$-2(\mathbf{I} - \mathbf{K}_k\mathbf{H}_k)\mathbf{P}_{k|k-1}\mathbf{H}_k^T + 2\mathbf{K}_k\mathbf{R}_k = 0. \tag{3.15}$$

Equation (3.15) can be solved for $\mathbf{K}_k$ as the following

$$\begin{aligned} \mathbf{K}_k\mathbf{R}_k &= \mathbf{P}_{k|k-1}\mathbf{H}_k^T - \mathbf{K}_k\mathbf{H}_k\mathbf{P}_{k|k-1}\mathbf{H}_k^T \\ \mathbf{P}_{k|k-1}\mathbf{H}_k^T &= \mathbf{K}_k(\mathbf{R}_k + \mathbf{H}_k\mathbf{P}_{k|k-1}\mathbf{H}_k^T) \\ \mathbf{K}_k &= \mathbf{P}_{k|k-1}\mathbf{H}_k^T(\mathbf{H}_k\mathbf{P}_{k|k-1}\mathbf{H}_k^T + \mathbf{R}_k)^{-1}. \end{aligned} \tag{3.16}$$

An alternative form of the updated covariance can be obtained by substituting Eq. (3.16) in Eq. (3.12) and this produces

$$\mathbf{P}_{k|k} = (\mathbf{I} - \mathbf{K}_k\mathbf{H}_k)\mathbf{P}_{k|k-1} \tag{3.17}$$

and

$$\mathbf{P}_{k|k}^{-1} = \mathbf{P}_{k|k-1}^{-1} + \mathbf{H}_k^T\mathbf{R}_k^{-1}\mathbf{H}_k. \tag{3.18}$$

The Kalman filter can now be summarised in Algorithm 3.1.

Kalman filter is one of the best tools available for linear state estimation. It provides optimal state estimation with minimum mean-squared error based on the system model, the input and corrupted output (measurement) data. The noises involved are assumed to be white noises. Its simple structure and straightforward design methodology make it popular with applications in almost every field of industry.

3.3 Kalman Information Filter

In some application areas, there are several sensors, rather than just one, used to measure the system output. That is, the measurements may come from different sets of sensors, so called *multiple sensing*. In such a situation, the employment of Kalman filters may not be effective neither direct, in comparison to another filtering method, the (Kalman) *Information Filter* (KIF). A Kalman information filter is an alternative, algebraic equivalent form of a Kalman filter. An information filter may be preferred over the standard Kalman filter due to its simpler update stage (Mutambara 1998).

In the information filtering process, the information state vector and the information matrix are introduced. The information state vector and the information matrix are propagated in the recursive process rather than the state vector and covariance matrix in the standard Kalman filter. The information matrix is the inverse of the

Algorithm 3.1 The Kalman filter (KF)

Initialise the state vector, $\widehat{\mathbf{x}}_{0|0}$, and the covariance, $\mathbf{P}_{0|0}$ (set $k = 1$).

Prediction

Evaluate the predicted state and covariance using

$$\begin{aligned}\widehat{\mathbf{x}}_{k|k-1} &= \mathbf{F}_{k-1}\hat{\mathbf{x}}_{k-1|k-1} + \mathbf{G}_{k-1}\mathbf{u}_{k-1}\\ \mathbf{P}_{k|k-1} &= \mathbf{F}_{k-1}\mathbf{P}_{k-1|k-1}\mathbf{F}_{k-1}^T + \mathbf{Q}_{k-1}.\end{aligned}$$

Measurement Update

Taking measurements $\mathbf{z}_k$

The updated state and covariance can be obtained as

$$\begin{aligned}\widehat{\mathbf{x}}_{k|k} &= \widehat{\mathbf{x}}_{k|k-1} + \mathbf{K}_k \nu_k\\ \mathbf{P}_{k|k} &= (\mathbf{I} - \mathbf{K}_k\mathbf{H}_k)\mathbf{P}_{k|k-1}\end{aligned}$$

where, the *innovation (residual) vector* ν_k and the *Kalman filter gain matrix* $\mathbf{K}_k$ are calculated by

$$\begin{aligned}\nu_k &= \mathbf{z}_k - \mathbf{H}_k\widehat{\mathbf{x}}_{k|k-1}\\ \mathbf{K}_k &= \mathbf{P}_{k|k-1}\mathbf{H}_k^T(\mathbf{H}_k\mathbf{P}_{k|k-1}\mathbf{H}_k^T + \mathbf{R}_k)^{-1}.\end{aligned}$$

Set $k = k + 1$ and go to the Prediction stage

Further, in the above, an *innovation covariance matrix* can be introduced as

$\S_k = \mathbf{H}_k\mathbf{P}_{k|k-1}\mathbf{H}_k^T + \mathbf{R}_k$

thus, the Kalman filter gain matrix can be calculated from

$\mathbf{K}_k = \mathbf{P}_{k|k-1}\mathbf{H}_k^T\S_k^{-1}$.

covariance matrix, and the information vector is the product of the information matrix and the state vector.

Consider the discrete process and measurement dynamics in (3.1) and (3.2). The process and measurement noises, $\mathbf{w}_{k-1}$ and $\mathbf{v}_k$, are again assumed to be Gaussian, and their corresponding covariance matrices are $\mathbf{Q}_{k-1}$ and $\mathbf{R}_k$, respectively. The prediction and update stages of the KIF are given below.

The predicted information state vector, $\hat{\mathbf{y}}_{k|k-1}$, and the predicted information matrix, $\mathbf{Y}_{k|k-1}$, are defined as

$$\begin{aligned}\hat{\mathbf{y}}_{k|k-1} &= \mathbf{Y}_{k|k-1}\hat{\mathbf{x}}_{k|k-1}\\ \mathbf{Y}_{k|k-1} &= \mathbf{P}_{k|k-1}{}^{-1}\\ &= \left[\mathbf{F}_{k-1}\mathbf{Y}_{k-1|k-1}{}^{-1}\mathbf{F}_{k-1}{}^T + \mathbf{Q}_{k-1}\right]^{-1}\end{aligned} \tag{3.19}$$

where $\mathbf{P}_{k|k-1}$ is the predicted covariance matrix and

$$\hat{\mathbf{x}}_{k|k-1} = \mathbf{F}_{k-1}\hat{\mathbf{x}}_{k-1|k-1} + \mathbf{G}_{k-1}\mathbf{u}_{k-1}. \tag{3.20}$$

The updated information state vector, $\hat{\mathbf{y}}_{k|k}$, and the updated information matrix, $\mathbf{Y}_{k|k}$, are

$$\hat{\mathbf{y}}_{k|k} = \hat{\mathbf{y}}_{k|k-1} + \mathbf{i}_k \tag{3.21}$$
$$\mathbf{Y}_{k|k} = \mathbf{Y}_{k|k-1} + \mathbf{I}_k. \tag{3.22}$$

The information state contribution, $\mathbf{i}_k$, and its associated information matrix, $\mathbf{I}_k$, are

$$\begin{aligned} \mathbf{i}_k &= \mathbf{H}_k{}^T \mathbf{R}_k{}^{-1} \left[\nu_k + \mathbf{H}_k \hat{\mathbf{x}}_{k|k-1} \right] \\ \mathbf{I}_k &= \mathbf{H}_k{}^T \mathbf{R}_k{}^{-1} \mathbf{H}_k \end{aligned} \tag{3.23}$$

where the measurement residual, ν_k, is

$$\nu_k = \mathbf{z}_k - \mathbf{H}_k(\hat{\mathbf{x}}_{k|k-1}). \tag{3.24}$$

The state vector and covariance matrix can be recovered by using MATLAB's *leftdivide* operator (Eustice et al. 2006)

$$\hat{\mathbf{x}}_{k|k} = \mathbf{Y}_{k|k} \backslash \hat{\mathbf{y}}_{k|k} \tag{3.25}$$
$$\mathbf{P}_{k|k} = \mathbf{Y}_{k|k} \backslash \mathbf{I}_n \tag{3.26}$$

where $\mathbf{I}_n$ is the state vector sized identity matrix.

Initialisation in the information filter is easier than in the Kalman filter and the update stage of information filter is computationally simpler than the Kalman filter. KIF (and indeed the information filter forms of other filters) can be shown to be more efficient than the Kalman filter in multiple sensor scenarios.

3.4 Discrete-Time $\mathcal{H}_\infty$ Filter

It is well understood that uncertainties are unavoidable in the operational environment of a real control system. The uncertainty can be classified into two categories: unwanted yet unknown external signals and dynamic perturbations. The former includes input and output disturbance (such as a gust on an aircraft), sensor noise and actuator noise, etc. The latter represents the discrepancy between the mathematical model and the actual dynamics of the system in operation. A mathematical model of any real system is always just an approximation of the true, physical reality of the system dynamics. Typical sources of the discrepancy include unmodelled (usually of high-frequency nature) dynamics, neglected nonlinearities in the modelling process, effects of deliberately reduced-order models, and system-parameter variations due to environmental changes and tear-and-wear factors. These modelling errors may adversely affect the stability and performance level of a control system. The Kalman filter discussed earlier assumes the process model has known dynamics and the noise sources have known statistics. However, due to the very possible scenarios of uncertainties, these assumptions may limit the application of Kalman filters in real applications. To overcome some of these limitations of Kalman filters, a few researchers have proposed $\mathcal{H}_\infty$ filters (Grimble and El Sayed 1990; Napgal and Khargonekar 1991; Yaesh and Shaked 1992; Shaked 1990; Basar 1991; Banavar and Speyer 1991; Park and Kailath 1995).

This section presents a brief introduction to an $\mathcal{H}_\infty$[4] filter which minimises the worst-case estimation error. This is in contrast to the Kalman filter which minimises the expected value of the variance of the estimation error. Furthermore, $\mathcal{H}_\infty$ filtering approach does not set any assumptions about the statistics of the process and measurement noises. For a detailed formulation and derivation please see for example Simon (2006), Shaked (1990), Einicke and White (1999) or Shen and Deng (1997).

The discrete-time process and observation models can be written as

$$\mathbf{x}_k = \mathbf{F}_{k-1}\mathbf{x}_{k-1} + \mathbf{G}_{k-1}\mathbf{u}_{k-1} + \mathbf{w}_{k-1}, \tag{3.27}$$

$$\mathbf{z}_k = \mathbf{H}_k\mathbf{x}_k + \mathbf{v}_k, \tag{3.28}$$

where k is the time index, $\mathbf{x}_k$ is the state vector, $\mathbf{u}_k$ the control input, $\mathbf{z}_k$ the measurement vector, and $\mathbf{w}_{k-1}$ and $\mathbf{v}_k$ are the process and measurement noises. These model equations are almost the same as that of the Kalman filter described in Sect. 3.2.1; except the assumption on noises. The noise terms $\mathbf{w}_{k-1}$ and $\mathbf{v}_k$ may be random with possibly unknown statistics, or even they may be deterministic. They may have a non-zero mean.

In a $\mathcal{H}_\infty$ filter, instead of directly estimating the state one may estimate a linear combination of the states

$$\mathbf{n}_k = \mathbf{L}_k\mathbf{x}_k. \tag{3.29}$$

Of course, by replacing $\mathbf{L}$ with the identity matrix, one can directly estimate the state vector.

In the game theory approach of $\mathcal{H}_\infty$ filtering, the performance measure is given by

$$\mathbf{J}_\infty = \frac{\sum_{k=1}^{N} \|\mathbf{n}_k - \widehat{\mathbf{n}}_k\|^2_{\mathbf{M}_k}}{\|\mathbf{x}_0 - \hat{\mathbf{x}}_0\|^2_{\mathbf{P}_0^{-1}} + \sum_{k=1}^{N}(\|\mathbf{w}_k\|^2_{\mathbf{Q}_k^{-1}} + \|\mathbf{v}_k\|^2_{\mathbf{R}_k^{-1}})} \tag{3.30}$$

where $\mathbf{P}_0$, $\mathbf{Q}_k$, $\mathbf{R}_k$, and $\mathbf{M}_k$ are symmetric, positive definite weighting matrices chosen by the user based on the problem at hand. The norm notation used in this section is $\|e\|^2_{S_k} = e^T S_k e$.

In this dynamic game theory framework, there are essentially two players: the designer and the nature. The designer's goal is to find the estimates of the error $\mathbf{n}_k - \widehat{\mathbf{n}}_k$ so that the cost $\mathbf{J}_\infty$ is minimised, while the Nature's goal is to maximise $\mathbf{J}_\infty$. The numerator of $\mathbf{J}_\infty$ can be viewed as the energy of the estimation error and

[4] $\mathcal{H}_\infty$ filters minimises the worst-case energy gain from the noise input to the estimation error, which is equivalent to minimising the corresponding $\mathcal{H}_\infty$ norm.

the denominator can be considered as the energy of the unknown disturbances. The Nature can simply put large magnitudes of $\mathbf{w}_k$, $\mathbf{v}_k$, and $\mathbf{x}_0$ to achieve its ultimate goal, and this makes the game unfair to the designer. Thus, the $\mathbf{J}_\infty$ is defined with $(\mathbf{x}_0 - \hat{\mathbf{x}}_0)$,$\mathbf{w}_k$, and $\mathbf{v}_k$ in the denominator. Then, the Nature needs to cleverly choose those disturbances in order to maximise $\mathbf{n}_k - \widehat{\mathbf{n}}_k$; likewise the designer also should be smart to find an estimation strategy to minimise $\mathbf{n}_k - \widehat{\mathbf{n}}_k$.

The task of the $\mathcal{H}_\infty$ filter is to minimise the state estimation error so that $\mathbf{J}_\infty$ is bounded by a prescribed threshold under the worst case $\mathbf{w}_k$, $\mathbf{v}_k$, and $\mathbf{x}_0$

$$\sup \mathbf{J}_\infty < \gamma^2 \tag{3.31}$$

where "sup" stands for *supremum*, $\gamma > 0$ is the error attenuation parameter.

Based on Eq. (3.31), the designer should find $\hat{\mathbf{x}}_k$ so that $\mathbf{J}_\infty < \gamma^2$ holds for any disturbances in $\mathbf{w}_k$, $\mathbf{v}_k$, and $\mathbf{x}_0$. The best the designer can do is to minimise $\mathbf{J}_\infty$ under worst-case disturbances, then the $\mathcal{H}_\infty$ filter can be interpreted as the following 'minmax' problem

$$\min_{\hat{\mathbf{x}}_k} \max_{\mathbf{w}_k, \mathbf{v}_k, \mathbf{x}_0} \mathbf{J}_\infty \tag{3.32}$$

For detailed analysis and solution procedure of the $\mathcal{H}_\infty$ filtering approach, please see Shen and Deng (1997) and Simon (2006). In this section, we use the $\mathcal{H}_\infty$ filter algorithm given in Yang et al. (2009), as the relevant equations in this reference are closely related to that of the Kalman filter.

The predicted state vector and auxiliary matrix of $\mathcal{H}_\infty$ filter are

$$\begin{aligned}
\widehat{\mathbf{x}}_{k|k-1} &= \mathbf{F}_{k-1}\hat{\mathbf{x}}_{k-1|k-1} + \mathbf{G}_{k-1}\mathbf{u}_{k-1} \\
\mathbf{P}_{k|k-1} &= \mathbf{F}_{k-1}\mathbf{P}_{k-1|k-1}\mathbf{F}_{k-1}^T + \mathbf{Q}_{k-1}
\end{aligned}$$

and the updated state and inverse of the updated auxiliary matrix can be obtained as

$$\hat{\mathbf{x}}_{k|k} = \hat{\mathbf{x}}_{k|k-1} + \mathbf{K}_\infty[\mathbf{z}_k - \mathbf{H}_k\hat{\mathbf{x}}_{k|k-1}] \tag{3.33}$$

$$\mathbf{P}_{k|k}^{-1} = \mathbf{P}_{k|k-1}^{-1} + \mathbf{H}_k^T\mathbf{R}_k^{-1}\mathbf{H}_k - \gamma^{-2}\mathbf{I}_n \tag{3.34}$$

where

$$\mathbf{K}_\infty = \mathbf{P}_{k|k-1}\mathbf{H}_k^T[\mathbf{H}_k\mathbf{P}_{k|k-1}\mathbf{H}_k^T + \mathbf{R}_k]^{-1} \tag{3.35}$$

and $\mathbf{I}_n$ denotes the identity matrix of dimension $n \times n$.

An $\mathcal{H}_\infty$ filter can be summarised in Algorithm 3.2, that has a structure similar to that of the Kalman filter.

It is interesting to note that, for very high values of γ, the updated auxiliary matrix of the $\mathcal{H}_\infty$ filter in Eq. (3.34) and the covariance matrix of the Kalman filter in Eq. (3.18) are equivalent. Hence, the $\mathcal{H}_\infty$ filter's performance can be matched with that of the Kalman filter, but the reverse is not true.

Algorithm 3.2 $\mathcal{H}_\infty$ Filter

Initialise the state vector, $\widehat{\mathbf{x}}_{0|0}$, and the auxiliary matrix, $\mathbf{P}_{0|0}$ (set $k = 1$).

Prediction

1: Evaluate the predicted state and auxiliary matrix using

$$\begin{aligned}\widehat{\mathbf{x}}_{k|k-1} &= \mathbf{F}_{k-1}\hat{\mathbf{x}}_{k-1|k-1} + \mathbf{G}_{k-1}\mathbf{u}_{k-1}\\ \mathbf{P}_{k|k-1} &= \mathbf{F}_{k-1}\mathbf{P}_{k-1|k-1}\mathbf{F}_{k-1}^T + \mathbf{Q}_{k-1}.\end{aligned}$$

Measurement Update

1: The updated state and the inverse of auxiliary matrix can be obtained as

$$\begin{aligned}\widehat{\mathbf{x}}_{k|k} &= \widehat{\mathbf{x}}_{k|k-1} + \mathbf{K}_\infty \nu_k\\ \mathbf{P}_{k|k}^{-1} &= \mathbf{P}_{k|k-1}^{-1} + \mathbf{H}_k^T\mathbf{R}_k^{-1}\mathbf{H}_k - \gamma^{-2}\mathbf{I}_n\end{aligned}$$

where,

$$\begin{aligned}\nu_k &= \mathbf{z}_k - \mathbf{H}_k\widehat{\mathbf{x}}_{k|k-1}\\ \mathbf{K}_\infty &= \mathbf{P}_{k|k-1}\mathbf{H}_k^T(\mathbf{H}_k\mathbf{P}_{k|k-1}\mathbf{H}_k^T + \mathbf{R}_k)^{-1}\end{aligned}$$

3.5 State Estimation and Control of a Quadruple-Tank System

In the previous sections, the Kalman and $\mathcal{H}_\infty$ filters were discussed, along with the information filter. This section deals with the combined state estimation and control problem for a quadruple-tank system using a Kalman filter, an information filter and an $\mathcal{H}_\infty$ filter. The controller design is based on the sliding mode control (SMC), which involves a stable sliding surface design followed by a control law design to ensure the system states onto the chosen surface (Bandyopadhyay et al. 2009). Combined SMC and state estimators for the quadruple-tank are then to be presented. Such approaches can be explored for many other practical systems. The basic structure of the approach is shown in Fig. 3.1. In this section, first, a brief literature survey on quadruple-tank system is given. The mathematical model of the quadruple-tank system is then presented, which is followed by SMC design and combined SMC-state estimators, and ended by the closed-loop simulations. Figures will be included to show the simulation results by using the Kalman filter and the $\mathcal{H}_\infty$ filter.

3.5.1 Quadruple-Tank System

The quadruple-tank considered here is an interesting multivariable plant consisting of four interconnected tanks and two pumps (Johansson 2000) that some researchers

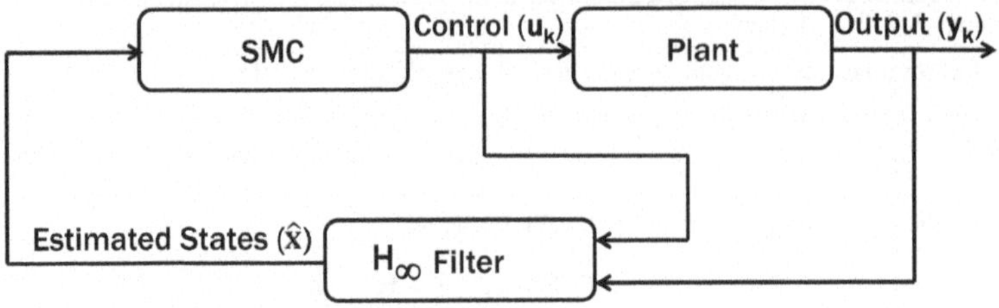

Fig. 3.1 A combined SMC-$\mathcal{H}_\infty$ filter approach

have used it to explore different control and estimation methods. Decentralised proportional-integral (PI) control, internal model control (IMC) and $\mathcal{H}_\infty$ controllers have all been designed for the quadruple-tank system (Vadigepalli et al. 2001). And, it has been reported that the IMC and $\mathcal{H}_\infty$ controllers provided better performance than the PI controller. Different nonlinear model predictive controllers were also proposed in Mercangöz and Doyle (2007) and Raff et al. (2006). Also, the interconnection and damping assignment passivity based control scheme for the quadruple-tank system was given in Johnsen and Allöwer (2007). Furthermore, a nonlinear sliding mode controller (SMC) with feedback linearisation was proposed and implemented in Biswas et al. (2009).

It is a well-known fact that all the states are required for a state feedback controller design. However, in most of real-world, practical applications, the states are not always available for feedback. Similarly, in the quadruple-tank system, only the first two states are assumed to be accessible for feedback and hence either one has to rely on output feedback control methods or the remaining two states are to be estimated for a state feedback design. The usage of an extended Kalman filter and high gain observers for the state estimation of a quadruple-tank system was given in Said and M'Sahli (2008) and state estimation in non-Gaussian domain using particle filter was described in Imtiaz et al. (2006). A very few researchers have demonstrated using combined controller-observer scheme for the quadruple-tank. Though in Okpanachi (2010), Kalman filter was used for the estimation of unavailable states of a quadruple-tank system and the controller was based on PID and IMC.

In this section, the discrete-time SMC using the Kalman and $\mathcal{H}_\infty$ filters, respectively, is to be designed for the quadruple-tank. Nonlinear sliding surface proposed in Bandyopadhyay et al. (2009) and Bandyopadhyay and Fulwani (2009) is considered to achieve better performance. The SMC for a quadruple-tank was also reported in Biswas et al. (2009). But, our approach is different from Biswas et al. (2009) in two aspects. In Biswas et al. (2009), the objective was to control just two states and linear sliding surfaces were constructed using only the first two states. However, in this section the main emphasis is to control all the four states and the nonlinear sliding surfaces are designed using complete state information. Secondly, SMC given in Theorem 1 of Biswas et al. (2009) requires the full state information. Whereas, we have assumed only two states are available from sensors and the complete state vector is estimated using Kalman and $\mathcal{H}_\infty$ filters, respectively.

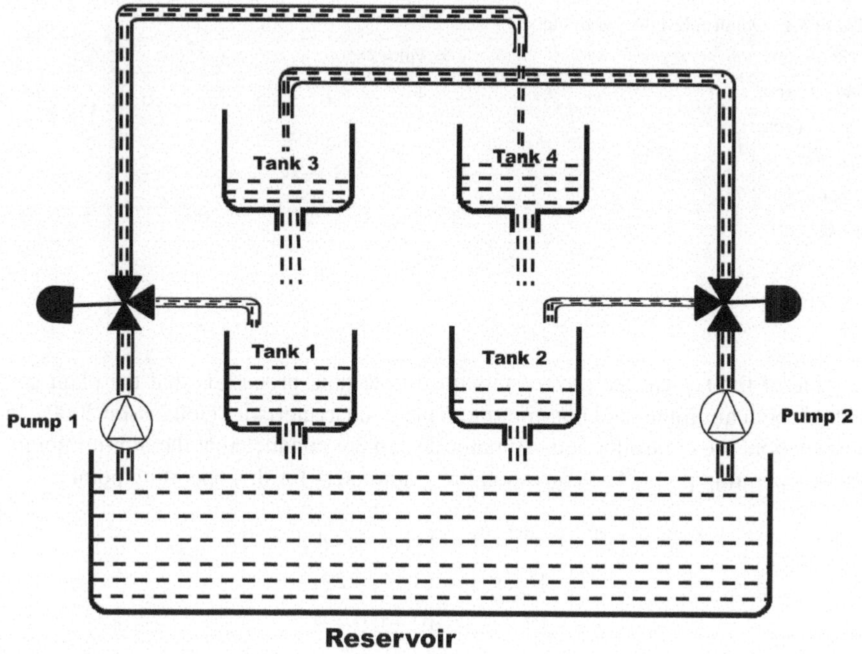

Fig. 3.2 Quadruple-Tank System (Johansson 2000)

The quadruple-tank system is shown in Fig. 3.2, which consists of four interconnected tanks, two pumps and two level sensors. For more details, see Johansson (2000) for the continuous-time plant model. The quadruple-tank is discretised using Euler's method with sampling time of $\Delta T = 0.1$ s. The discrete-time nonlinear quadruple-tank model is

$$\mathbf{x}_{k+1} = \begin{bmatrix} x_{1k} + \Delta T(-\frac{a_1}{A_1}\sqrt{2gx_{1k}} + \frac{a_3}{A_1}\sqrt{2gx_{3k}} + \frac{\gamma_1 k_1}{A_1}u_1) \\ x_{2k} + \Delta T(-\frac{a_2}{A_2}\sqrt{2g_{2k}} + \frac{a_4}{A_2}\sqrt{2gx_{4k}} + \frac{\gamma_2 k_2}{A_2}u_2) \\ x_{3k} + \Delta T(-\frac{a_3}{A_3}\sqrt{2gx_{2k}} + \frac{(1-\gamma_2)k_2}{A_3}u_2) \\ x_{4k} + \Delta T(-\frac{a_4}{A_4}\sqrt{2gx_{4k}} + \frac{(1-\gamma_1)k_1}{A_4}u_1) \end{bmatrix}$$

where, the state vector $\mathbf{x}$, consisting of water levels of all tanks, is $\mathbf{x} = [x_1, x_2, x_3, x_4]^T$. The inputs are the voltages to the two pumps, $[u_1, u_2]^T$ and the outputs are the voltages from the level measurements of the first two tanks. A_i is the cross section of Tank i and a_i is the cross section of outlet hole. The first input u_1 directly affects the first and fourth states, whereas the second input u_2 has direct influence on the second and third states. The outputs are the measured level signals, $k_c x_1$ and $k_c x_2$. The additive process and sensor noises are added to the state and output vectors, respectively. The parameter values used for simulation are given in Table 3.1.

Table 3.1 Quadruple-tank parameters

Parameters	Values
A_1, A_3 (cm^2)	28
A_2, A_4 (cm^2)	32
a_1, a_3 (cm^2)	0.071
a_2, a_4 (cm^2)	0.057
k_c (V/cm)	0.5
g (cm/s^2)	981

One of the typical features of this quadruple-tank process is that the plant can exhibit both minimum and non-minimum phase characteristics (Johansson 2000). In this section, the controller and estimator design are conducted at the non-minimum phase operating point. The corresponding parameters for this operating point are

$$\begin{aligned}[x_1^0, x_2^0, x_3^0, x_4^0] &= [12.6, 13, 4.8, 4.9] \\ [k_1, k_2] &= [3.14, 3.29] \\ [\gamma_1, \gamma_2] &= [0.43, 0.34]\end{aligned}$$

The quadruple-tank is linearised at the above operating point and the mathematical model is given below

$$\mathbf{x}_{k+1} = \mathbf{F}\mathbf{x}_k + \mathbf{G}\mathbf{u}_k + \mathbf{w}_k \tag{3.36}$$

$$\mathbf{z}_k = \mathbf{H} + \mathbf{v}_k \tag{3.37}$$

where, the state space matrices are

$$\mathbf{F} = \begin{bmatrix} 1 + \frac{a_{11}\Delta T}{2\sqrt{x_1^0}} & 0 & \frac{a_{13}\Delta T}{2\sqrt{x_3^0}} & 0 \\ 0 & 1 + \frac{a_{21}\Delta T}{2\sqrt{x_2^0}} & 0 & \frac{a_{24}\Delta T}{2\sqrt{x_4^0}} \\ 0 & 0 & 1 + \frac{a_{31}\Delta T}{2\sqrt{x_3^0}} & 0 \\ 0 & 0 & 0 & 1 + \frac{a_{41}\Delta T}{2\sqrt{x_4^0}} \end{bmatrix}$$

$$\mathbf{G} = \begin{bmatrix} b_1\Delta T & 0 \\ 0 & b_2\Delta T \\ 0 & b_3\Delta T \\ b_4\Delta T & 0 \end{bmatrix}$$

$$\mathbf{H} = \begin{bmatrix} k_c & 0 & 0 & 0 \\ 0 & k_c & 0 & 0 \end{bmatrix}$$

and the parameters are
$a_{11} = -\frac{a_1}{A_1}\sqrt{2g}$, $a_{12} = \frac{a_3}{A_1}\sqrt{2g}$, $a_{21} = -\frac{a_2}{A_2}\sqrt{2g}$, $a_{24} = \frac{a_4}{A_2}\sqrt{2g}$, $a_{31} = -\frac{a_3}{A_3}\sqrt{2g}$, $a_{41} = -\frac{a_4}{A_4}\sqrt{2g}$, $b_1 = \frac{\gamma_1 k_1}{A_1}$, $b_2 = \frac{\gamma_2 k_2}{A_2}$, $b_3 = \frac{(1-\gamma_2)k_2}{A_3}$, and $b_4 = \frac{(1-\gamma_1)k_1}{A_1}$.

3.5.2 *Sliding Mode Control of the Quadruple-Tank System*

This section deals with the SMC of the quadruple-tank system. The objective is to control all the four states, unlike in Biswas et al. (2009), where only two states are controlled. For an improved performance, nonlinear sliding surfaces are considered during SMC design. Using a nonlinear sliding surface, the damping ratio of a system can be varied from its initial, low value to a final, high value. The initial low damping ratio results in a faster response and the later high damping avoids the possible overshoot. Thus the nonlinear surface ascertains the reduction in settling time without any overshoot (Bandyopadhyay et al. 2009; Bandyopadhyay and Fulwani 2009).

SMC Design

The plant in Eq. (3.36) is not in the regular form, as the input matrix **G** does not have any zero-row vectors. Before designing the SMC control, the plant need to be transformed into the regular form. The plant can be transformed into the regular form by using the transformation matrix

$$T = \begin{bmatrix} -\frac{1}{b_1} & 0 & 0 & \frac{1}{b_4} \\ 0 & -\frac{1}{b_2} & \frac{1}{b_3} & 0 \\ 0 & 0 & \frac{1}{b_3} & 0 \\ 0 & 0 & 0 & \frac{1}{b_4} \end{bmatrix}.$$

The new transformed system can be written as

$$\mathbf{y}_{k+1} = (T\mathbf{F}T^{-1})\mathbf{y}_k + T\mathbf{G}\mathbf{u}_k + T\mathbf{G}\mathbf{w}_k \tag{3.38}$$

where $\mathbf{y}_k = T\mathbf{x}_k$ and the system can be further expressed as

$$\mathbf{y}_{k+1} = \begin{bmatrix} \mathbf{y}_{u,k+1} \\ \mathbf{y}_{l,k+1} \end{bmatrix} = \begin{bmatrix} \mathbf{y}_{11} & \mathbf{y}_{12} \\ \mathbf{y}_{21} & \mathbf{y}_{22} \end{bmatrix} \begin{bmatrix} \mathbf{y}_{u,k} \\ \mathbf{y}_{l,k} \end{bmatrix} + \begin{bmatrix} \mathbf{0} \\ \mathbf{G}_l \end{bmatrix} \mathbf{u}_k + \begin{bmatrix} \mathbf{0} \\ \mathbf{G}_l \end{bmatrix} \mathbf{w}_k \tag{3.39}$$

where

$$\mathbf{y}_{11} = \begin{bmatrix} 1 + \frac{a_{11}\Delta T}{2\sqrt{x_1^0}} & 0 \\ 0 & 1 + \frac{a_{21}\Delta T}{2\sqrt{x_2^0}} \end{bmatrix}$$

$$\mathbf{y}_{12} = \begin{bmatrix} \frac{a_{13}b_3\Delta T}{2b_1\sqrt{x_3^0}} & \frac{\Delta T}{2}\left(\frac{a_{41}}{\sqrt{x_4^0}} - \frac{a_{11}}{\sqrt{x_1^0}}\right) \\ \frac{\Delta T}{2}\left(\frac{a_{31}}{\sqrt{x_3^0}} - \frac{a_{21}}{\sqrt{x_2^0}}\right) & \frac{a_{24}b_4\Delta T}{2b_2\sqrt{x_4^0}} \end{bmatrix}$$

$$\mathbf{y}_{21} = \begin{bmatrix} 0 & 0 \\ 0 & 0 \end{bmatrix}$$

$$\mathbf{y}_{22} = \begin{bmatrix} 1 + \frac{a_{31}\Delta T}{2\sqrt{x_3^0}} & 0 \\ 0 & 1 + \frac{a_{41}\Delta T}{2\sqrt{x_4^0}} \end{bmatrix}$$

$$\mathbf{G}_l = \begin{bmatrix} 0 & \Delta T \\ \Delta T & 0 \end{bmatrix}.$$

Nonlinear Sliding Surface Design

The performance of the SMC-based, closed-loop system can be enhanced by using the nonlinear surfaces (Bandyopadhyay et al. 2009; Bandyopadhyay and Fulwani 2009) and hence in the present book a nonlinear sliding surface is to be adopted.

Let the nonlinear sliding surface (Bandyopadhyay et al. 2009) be

$$s_k = c_k^T \mathbf{y}_k \tag{3.40}$$

where

$$c_k^T = [K - \psi(\mathbf{z}_k)\mathbf{y}_{12}^T P(\mathbf{y}_{11} - \mathbf{y}_{12}K) \quad I_2] \tag{3.41}$$

and I_2 is the second-order identity matrix.
The gain matrix,

$$K = \begin{bmatrix} -0.66651 & -0.2026 \\ 0.06417 & -0.6641 \end{bmatrix} \tag{3.42}$$

is obtained by solving the linear quadratic regulator (LQR) problem for $\mathbf{y}_{11}$ and $\mathbf{y}_{12}$, such that $(\mathbf{y}_{11} - \mathbf{y}_{12}K)$ have stable eigenvalues (i.e. in the open, left-half complex plane). The weighting matrices considered for the LQR design are chosen as

$$Q_{lqr} = \begin{bmatrix} 10 & 0 \\ 0 & 10 \end{bmatrix} \, and \quad R_{lqr} = \begin{bmatrix} 10 & 0 \\ 0 & 10 \end{bmatrix}.$$

The matrix P, required for Eq. (3.41) is obtained by solving the following Lyapunov equation:

$$P = (\mathbf{y}_{11} - \mathbf{y}_{12}K)^T P(\mathbf{y}_{11} - \mathbf{y}_{12}K) + W. \tag{3.43}$$

By selecting W as I_2, the corresponding P matrix is

$$\begin{bmatrix} 1223.39 & -32.47 \\ -32.57 & 162.17 \end{bmatrix}.$$

The nonlinear function, $\psi(\mathbf{z}_k)$, can be chosen as Bandyopadhyay et al. (2009)

$$\psi(\mathbf{z}_k) = \begin{bmatrix} -\beta_1 e^{-m_1|x^0_{1,k-1}|} & 0 \\ 0 & -\beta_2 e^{-m_2|x^0_{2,k-1}|} \end{bmatrix} \tag{3.44}$$

where β_1, β_2, m_1 and m_2 are the tuning parameters. These parameters are selected as unity and $\psi(\mathbf{z}_k)$ is chosen such that it satisfies the below condition (Bandyopadhyay and Fulwani 2009; Cheng and Peng 2007)

$$2\psi(\mathbf{z}_k) + \psi(\mathbf{z}_k)\mathbf{y}_{12}^T P \mathbf{y}_{12}\psi(\mathbf{z}_k) \leq 0. \tag{3.45}$$

Control Law

Now the control law will be derived. From Eq. (3.40), s_{k+1} can be written as

$$\begin{aligned} s_{k+1} &= c_{k+1}^T \mathbf{y}_{k+1} \\ s_{k+1} &= c_{k+1}^T T \mathbf{x}_{k+1}. \end{aligned} \tag{3.46}$$

By using Eqs. (3.36) and (3.46), s_{k+1} can be written as

$$s_{k+1} = c_{k+1}^T T\mathbf{F}\mathbf{x}_k + c_{k+1}^T T\mathbf{G}\mathbf{u}_k + c_{k+1}^T T\mathbf{G}\mathbf{w}_k. \tag{3.47}$$

In the discrete-time sliding mode control, one of the objectives is to achieve the sliding surfaces, $s_k = 0$ in a finite time (Bandyopadhyay et al. 2009). This can be achieved by an equivalent control law (Utkin et al. 2009) by setting

$$s_{k+1} = 0. \tag{3.48}$$

By using Eqs. (3.47) and (3.48), the control law can be derived as

$$c_{k+1}^T T\mathbf{F}\mathbf{x}_k + c_{k+1}^T T\mathbf{G}\mathbf{u}_k + c_{k+1}^T T\mathbf{G}\mathbf{w}_k = 0 \tag{3.49}$$

and

$$\mathbf{u}_k = -(c_{k+1}^T T\mathbf{G})^{-1}(c_{k+1}^T T\mathbf{F}\mathbf{x}_k + c_{k+1}^T T\mathbf{G}\mathbf{w}_k). \tag{3.50}$$

The aforementioned control law contains uncertain terms or process noises. In general these uncertain terms are unknown. They can however be replaced by the average of known bounds of these terms (Bandyopadhyay and Fulwani 2009). The modified control law can be written as

$$\mathbf{u}_k = -(c_{k+1}^T T\mathbf{G})^{-1}(c_{k+1}^T T\mathbf{F}\mathbf{x}_k + d_m). \tag{3.51}$$

where d_m is the average of the lower and upper bounds of $c_{k+1}^T T\mathbf{G}\mathbf{w}_k$.

From Eq. (3.51), it can be seen that the control law at the kth time instant requires the information at $k+1$th time instant, and in general this is not feasible. However, from Eqs. (3.41) and (3.44), only the output information at kth instant ($x_{1,k}$ and $x_{2,k}$) is required to find c_{k+1} and hence the control input, $\mathbf{u}_k$, can be evaluated using just the output information at kth instant.

Stability of the Nonlinear Sliding Surface

From Eq. (3.39),

$$\mathbf{y}_{u,k+1} = \mathbf{y}_{11}\mathbf{y}_{u,k} + \mathbf{y}_{12}\mathbf{y}_{l,k} \tag{3.52}$$

During the sliding phase, $s_k = 0$, and hence Eq. (3.40) can be written as

$$c_k^T \mathbf{y}_k = 0 \tag{3.53}$$

$$\Rightarrow \quad \left[K - \psi(\mathbf{z}_k)\mathbf{y}_{12}^T P \mathbf{y}_{eq} \quad I_2\right]\left[\mathbf{y}_{u,k} \quad \mathbf{y}_{l_k}\right]^T = 0 \tag{3.54}$$

$$\Rightarrow \quad \mathbf{y}_{l,k} = -[K - \psi(\mathbf{z}_k)\mathbf{y}_{12}^T P \mathbf{y}_{eq}]\mathbf{y}_{u,k} \tag{3.55}$$

where $\mathbf{y}_{eq} = \mathbf{y}_{11} - \mathbf{y}_{12}K$.

From Eqs. (3.52) and (3.55),

$$\mathbf{y}_{u,k+1} = \mathbf{y}_{11}\mathbf{y}_{u,k} + \mathbf{y}_{12}(-[K - \psi(\mathbf{z}_k)\mathbf{y}_{12}^T P \mathbf{y}_{eq}])\mathbf{y}_{u,k} \tag{3.56}$$

$$= (\mathbf{y}_{11} - \mathbf{y}_{12}K)\mathbf{y}_{u,k} + \mathbf{y}_{12}\psi(\mathbf{z}_k)\mathbf{y}_{12}^T P \mathbf{y}_{eq}\mathbf{y}_{u,k} \tag{3.57}$$

$$= (\mathbf{y}_{12}\psi(\mathbf{z}_k)\mathbf{y}_{12}^T P + I)\mathbf{y}_{eq}\mathbf{y}_{u,k}. \tag{3.58}$$

The stability of the nonlinear sliding surface can be proved by using the Lyapunov theory. It is assumed that the Lyapunov function for the system defined in Eq. (3.58) is

$$V_k = \mathbf{y}_{u,k}^T P \mathbf{y}_{u,k}. \tag{3.59}$$

The increment of V_k is

$$\begin{aligned}
\triangle V_k &= V_{k+1} - V_k \qquad (3.60)\\
&= \mathbf{y}_{u,k+1}^T P \mathbf{y}_{u,k+1} - \mathbf{y}_{u,k}^T P \mathbf{y}_{u,k}\\
&= (\mathbf{y}_{u,k}^T \mathbf{y}_{eq}^T + \mathbf{y}_{u,k}^T \mathbf{y}_{eq}^T P \mathbf{y}_{12} \psi(\mathbf{z}_k) \mathbf{y}_{12}^T) P (\mathbf{y}_{eq} \mathbf{y}_{u,k} + \mathbf{y}_{12} \psi(\mathbf{z}_k) \mathbf{y}_{12}^T P \mathbf{y}_{eq} \mathbf{y}_{u,k}) - \mathbf{y}_{u,k}^T P \mathbf{y}_{u,k}\\
&= \mathbf{y}_{u,k}^T \mathbf{y}_{eq}^T P \mathbf{y}_{eq} \mathbf{y}_{u,k} + \mathbf{y}_{u,k}^T \mathbf{y}_{eq}^T P \mathbf{y}_{12} \psi(\mathbf{z}_k) \mathbf{y}_{12}^T P \mathbf{y}_{eq} \mathbf{y}_{u,k} + \mathbf{y}_{u,k}^T \mathbf{y}_{eq}^T P \mathbf{y}_{12} \psi(\mathbf{z}_k) \mathbf{y}_{12}^T P \mathbf{y}_{eq} \mathbf{y}_{u,k} +\\
&\quad \mathbf{y}_{u,k}^T \mathbf{y}_{eq}^T P \mathbf{y}_{12} \psi(\mathbf{z}_k) \mathbf{y}_{12}^T P \mathbf{y}_{12} \psi(\mathbf{z}_k) \mathbf{y}_{12}^T P \mathbf{y}_{eq} \mathbf{y}_{u,k} - \mathbf{y}_{u,k}^T P \mathbf{y}_{u,k}\\
&= -\mathbf{y}_{u,k}^T (P - \mathbf{y}_{eq}^T P \mathbf{y}_{eq}) \mathbf{y}_{u,k} + \mathbf{y}_{u,k}^T \mathbf{y}_{eq}^T P \mathbf{y}_{12} [2\psi(\mathbf{z}_k) + \psi(\mathbf{z}_k) \mathbf{y}_{12}^T P \mathbf{y}_{12} \psi(\mathbf{z}_k)] \mathbf{y}_{12}^T P \mathbf{y}_{eq} \mathbf{y}_{u,k}\\
&= -\mathbf{y}_{u,k}^T W \mathbf{y}_{u,k} + M^T [2\psi(\mathbf{z}_k) + \psi(\mathbf{z}_k) \mathbf{y}_{12}^T P \mathbf{y}_{12} \psi(\mathbf{z}_k)] M \qquad (3.61)
\end{aligned}$$

where $M = \mathbf{y}_{u,k}^T \mathbf{y}_{eq}^T P \mathbf{y}_{12}$.

From Eqs. (3.45) and (3.61), one can write

$$\nabla V \leq -\mathbf{y}_{u,k}^T W \mathbf{y}_{u,k}. \qquad (3.62)$$

Since the increment of the Lyapunov function is of negative definite, the equilibrium point for Eq. (3.58) is stable, and hence the designed nonlinear sliding surface is stable. In a similar way by constructing the Lyapunov function of the sliding surface, s_{k+1}, the increment of the Lyapunov function can be easily shown to be negative definite which in turn proves the existence of the sliding mode (Bandyopadhyay and Fulwani 2009).

3.5.3 Combined Schemes: Simulations and Results

By using the filter algorithms described in Sects. 3.2 and 3.4, all the four states of the quadruple-tank can be estimated by using x_1 and x_2. One may note that, the control input **u** required for the quadruple-tank is obtained from SMC controller given in Sect. 3.5.2. This proposed scheme for the quadruple-tank system is shown in Fig. 3.3. The process noise is added to all the four states, whereas the measurement noise is added to the last two states only; the quadruple-tank block in Fig. 3.3 assumed to have additive noises and hence separate noises are not added in the block diagram. The usage of the $\mathcal{H}_\infty$ filter in this work is required for two purposes; it is used to estimate the two unavailable states, and to inherently filter out the process and sensor noises. One may note that in this work although the first two states of the quadruple-tank are available from the sensors, we are still using the full-order state estimation rather than a reduced-order state estimation. The first two estimated states are less noisy than actual measured states and are beneficial for the state feedback SMC control design.

This section describes the various simulations done for closed-loop quadruple-tank system. Although, the controller and estimator designs are conducted for the linearised model, the simulations in this section are performed on the full nonlinear

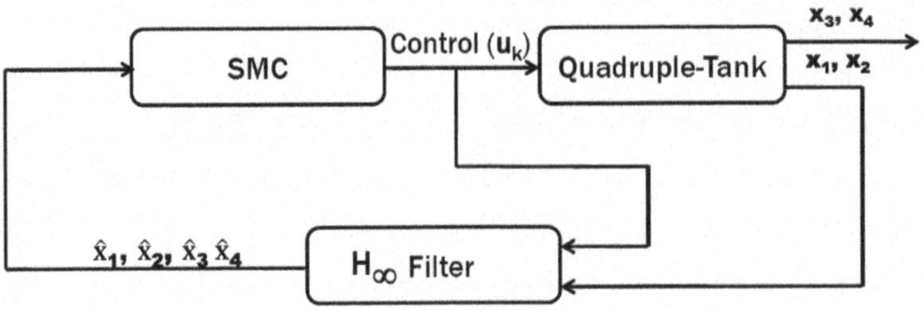

Fig. 3.3 Proposed scheme using SMC and $\mathcal{H}_\infty$ filter for quadruple-tank system

model. The SMC given in Sect. 3.5.2 and the estimator in Sects. 3.2 and 3.4 are considered in the simulations. The first two sensed states from quadruple-tank are given to the filters, which then estimates all the four states. These estimated states from estimators are used by the SMC, which then provides the input to the quadruple-tank system. The initial values of the plant are perturbed by $+15\%$ of their nominal values, and the objective is to bring back the perturbed states to the actual initial conditions. The chosen initial covariance matrix, $\mathbf{P}_{0|0}$, is

$$\mathbf{P}_{0|0} = \begin{bmatrix} 1\,0\,0\,0 \\ 0\,1\,0\,0 \\ 0\,0\,1\,0 \\ 0\,0\,0\,1 \end{bmatrix}.$$

Two sets of simulations have been performed to show the efficacy of the proposed method and to be compared with the Kalman filter's response. The first set involves the closed-loop simulation in the presence of Gaussian noises and the second one involves the simulation with non-Gaussian noises.

Simulation in the Presence of Gaussian Noises

In this subsection, it is assumed that the plant and sensor noises, $\mathbf{w}_k$ and $\mathbf{v}_k$, are both zero-mean Gaussian. The standard deviations for all the four states and the measurements are 0.0316. The corresponding covariance matrices for the Kalman filter are

$$\mathbf{Q} = \begin{bmatrix} 0.001 & 0 & 0 & 0 \\ 0 & 0.001 & 0 & 0 \\ 0 & 0 & 0.001 & 0 \\ 0 & 0 & 0 & 0.001 \end{bmatrix}$$

$$\mathbf{R} = \begin{bmatrix} 0.001 & 0 \\ 0 & 0.001 \end{bmatrix}.$$

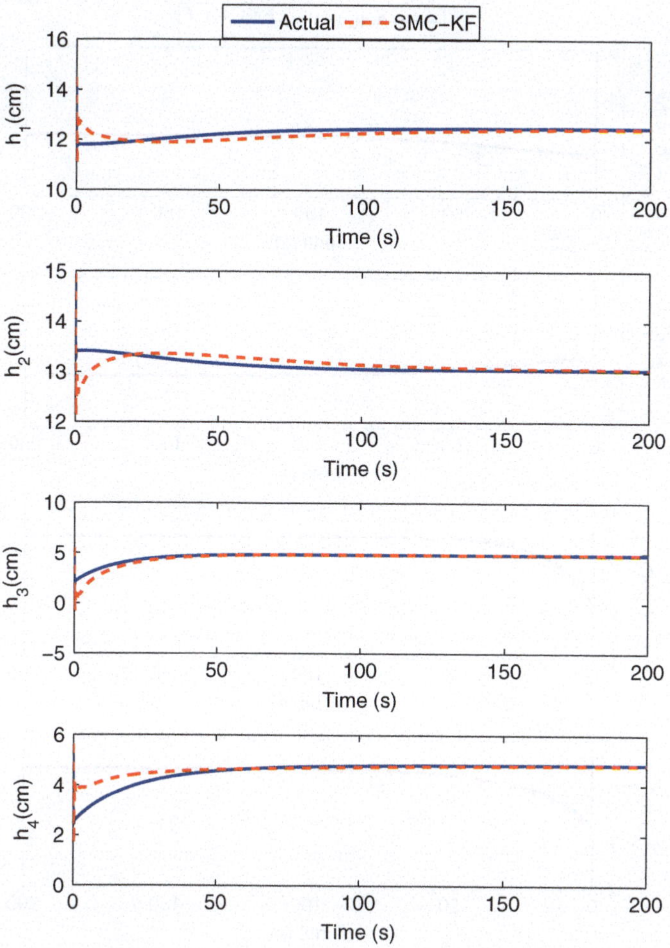

Fig. 3.4 Actual and estimated states of the quadruple-tank using the Kalman filter in the presence of Gaussian noises

The tuning parameters for the $\mathcal{H}_\infty$ filter simulations are

$$\mathbf{Q} = \begin{bmatrix} 0.01 & 0 & 0 & 0 \\ 0 & 0.01 & 0 & 0 \\ 0 & 0 & 0.01 & 0 \\ 0 & 0 & 0 & 0.01 \end{bmatrix}$$

$$\mathbf{R} = \begin{bmatrix} 0.01 & 0 \\ 0 & 0.01 \end{bmatrix}.$$

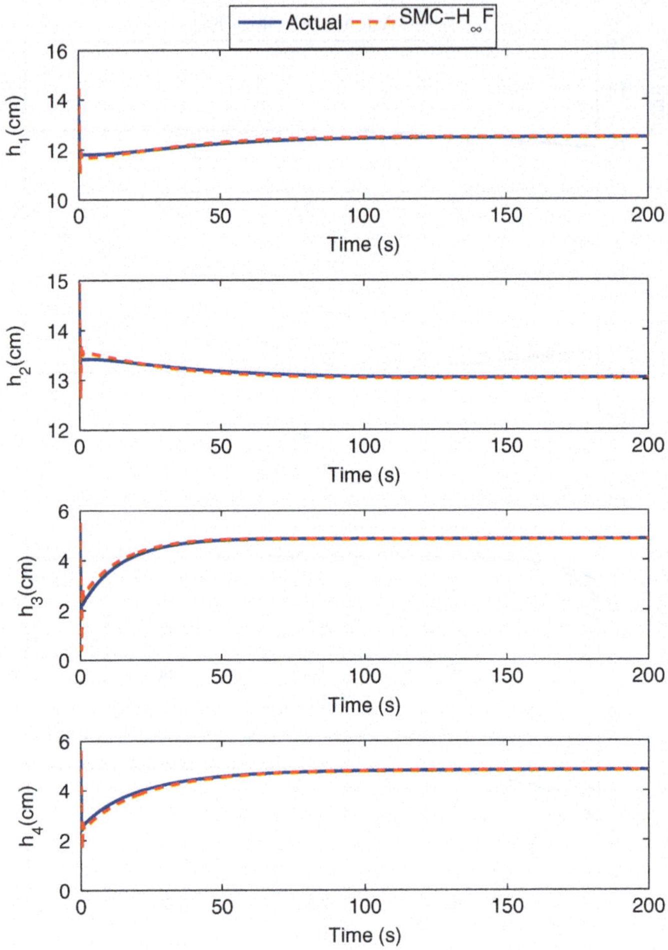

Fig. 3.5 Actual and estimated states of the quadruple-tank using $\mathcal{H}_\infty$ filter in the presence of Gaussian noises

and the performance bound, γ, is chosen as 1. These tuning parameters are selected by the trial-and-error method. One can note that, the chosen standard deviations for process and measurement noises using $\mathcal{H}_\infty$ filter are higher than the Kalman filter. The closed-loop performance can further be improved by using rigorous tuning methods; at the cost of increased computation complexity.

The SMC-based quadruple-tank levels using the Kalman filter are shown in Fig. 3.4, whereas for the $\mathcal{H}_\infty$ filter are shown in Fig. 3.5. In both figures, the actual and estimated perturbed states reaches their actual values in the finite time. The estimated states closely follows the actual states and the error between them decreases with time. From the initial transient response, it can be seen that the $\mathcal{H}_\infty$ filter's response is faster than that of the Kalman filter.

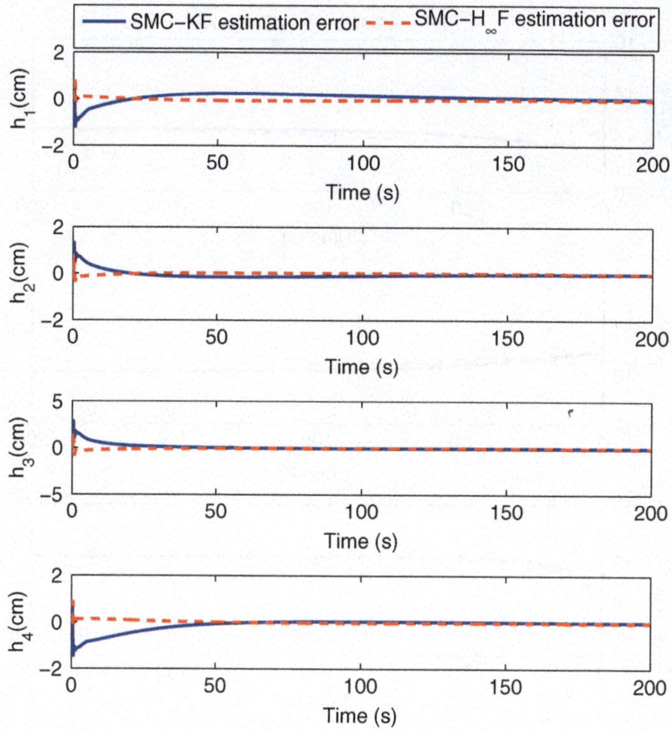

Fig. 3.6 Estimation errors for the quadruple-tank using Kalman and $\mathcal{H}_\infty$ filters in the presence of Gaussian noises

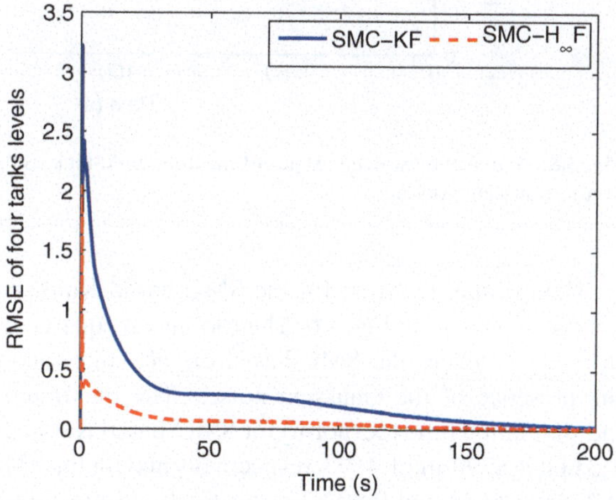

Fig. 3.7 RMSEs using Kalman and $\mathcal{H}_\infty$ filters in the presence of non-Gaussian noises

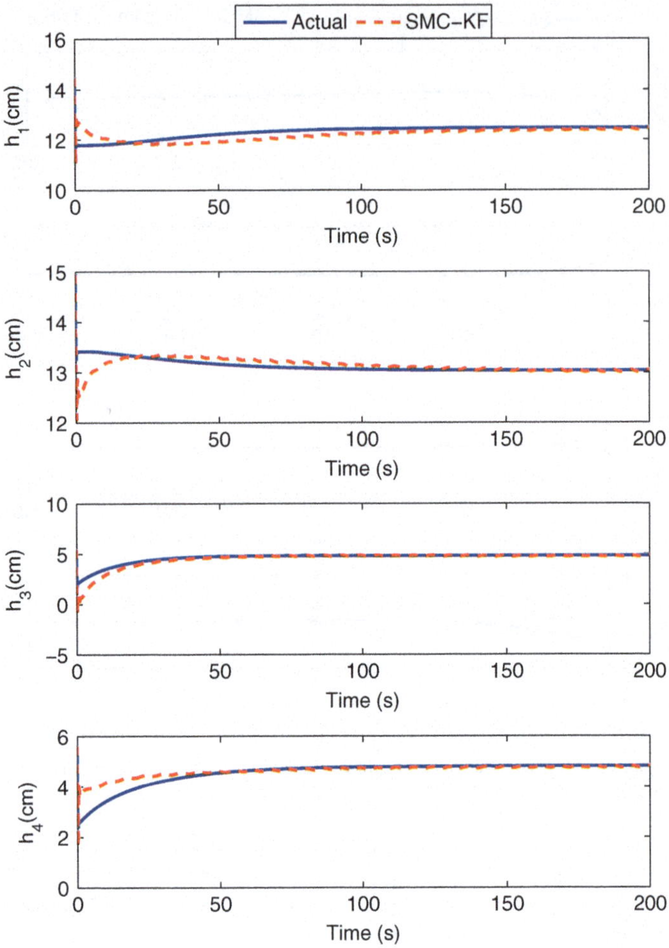

Fig. 3.8 Actual and estimated states of the quadruple-tank using the Kalman filter in the presence of non-Gaussian noises

The estimation errors for the SMC-based Kalman and $\mathcal{H}_\infty$ filters for Gaussian noises are shown in Fig. 3.6. The root mean-squared error (RMSE) plots are shown in Fig. 3.7, where the SMC-based on $\mathcal{H}_\infty$ filter shows the better performance in the presence of the Gaussian noises. The maximum state estimation errors over the simulation time (∞-norm) for SMC-based Kalman filter's four states are 1.2051, 1.3111, 2.8799 and 1.4693, respectively and for the SMC-based $\mathcal{H}_\infty$ filter are 0.7185, 0.7254, 1.6648 and 0.8921, respectively. In the simulations, the quadruple-tank is excited by Gaussian noises for both the Kalman and $\mathcal{H}_\infty$ filters. In the Kalman filter, if the standard deviation of noises once fixed then the corresponding covariances are the square of the standard deviations. However, in the $\mathcal{H}_\infty$ filter, there is a freedom to select tuning parameters irrespective of the noises. Both Kalman and $\mathcal{H}_\infty$ filters'

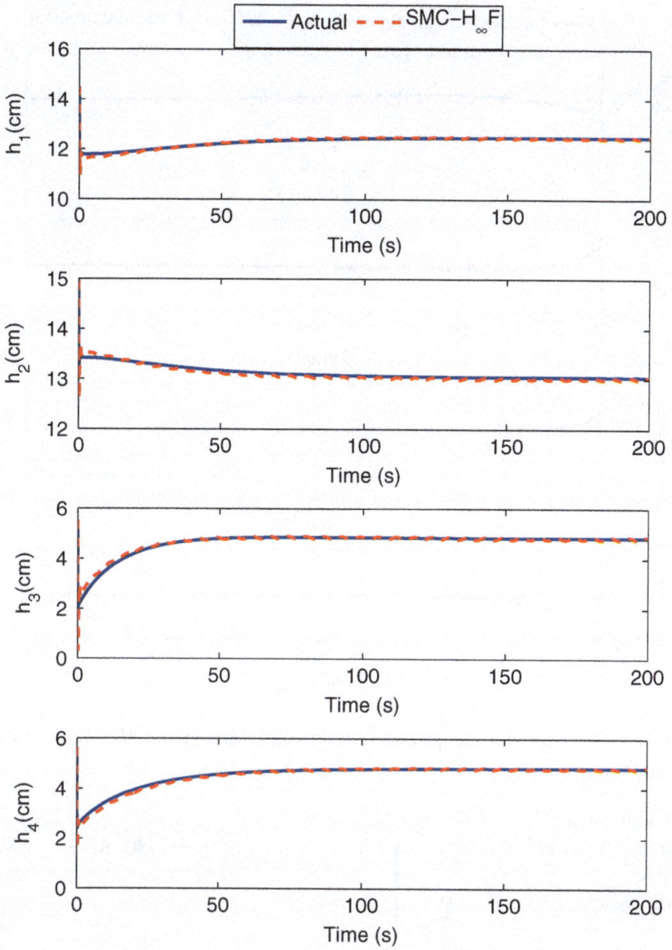

Fig. 3.9 Actual and estimated states of the quadruple-tank using $\mathcal{H}_\infty$ filter in the presence of non-Gaussian noises

performances can further improved by tuning **Q** and **R**. The results in this section are shown for the full nonlinear model. When the same simulations are repeated with linear open-loop plant model, the Kalman filter's response is better than the $\mathcal{H}_\infty$ filter's response, as the Kalman filter is the optimal estimator for linear-Gaussian systems.

Simulation in the Presence of Non-Gaussian Noises

In most of real-life applications, the assumption of zero-mean and Gaussian noises may not be valid. To validate the proposed approach, non-zero-mean and non-

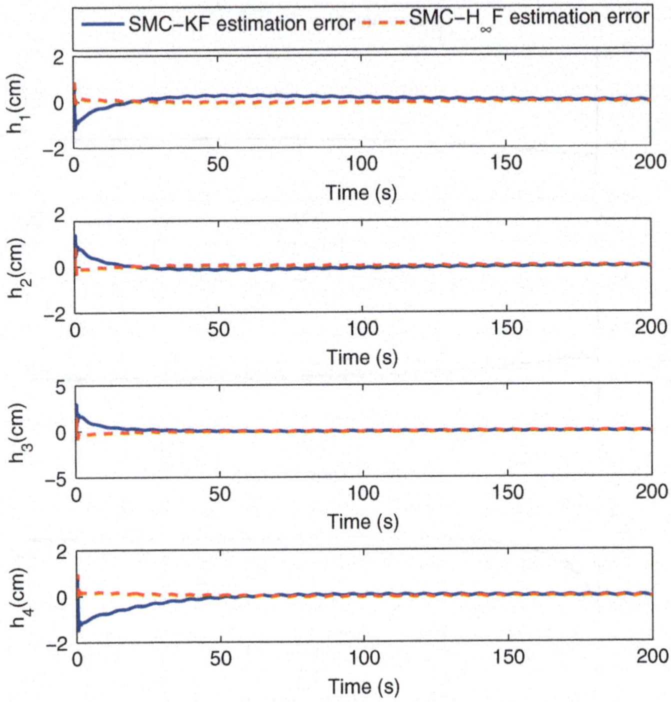

Fig. 3.10 Estimation errors for the quadruple-tank using Kalman and $\mathcal{H}_\infty$ filters in the presence of non-Gaussian noises

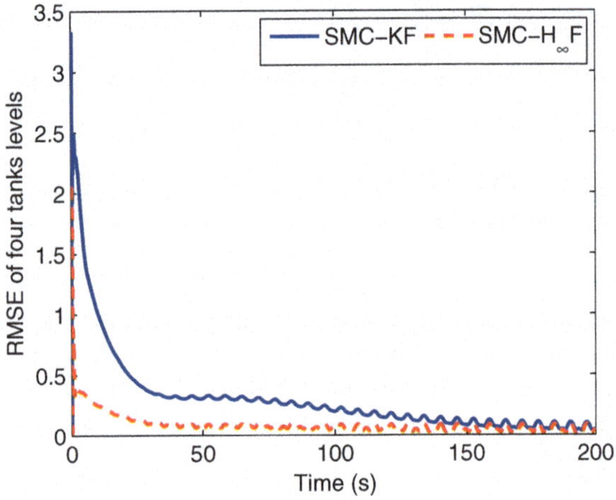

Fig. 3.11 RMSEs using Kalman and $\mathcal{H}_\infty$ filters in the presence of non-Gaussian noises

Gaussian noises are considered for the simulations. The process and measurement noises are

$$\mathbf{w}_k = \begin{bmatrix} 0.01 + 0.01 \times sin(0.1k) \\ 0.01 + 0.01 \times sin(0.1k) \\ 0.01 + 0.01 \times sin(0.1k) \\ 0.01 + 0.01 \times sin(0.1k) \end{bmatrix}$$

$$\mathbf{v}_k = \begin{bmatrix} 0.01 + 0.01 \times sin(0.1k) \\ 0.01 + 0.01 \times sin(0.1k) \end{bmatrix}.$$

The maximum magnitudes of $\mathbf{w}_k$ and $\mathbf{v}_k$ are used in the noise covariance matrices for the Kalman filter. For the $\mathcal{H}_\infty$ filter, the tuning parameters given in the simulation with Gaussian noises are considered. The closed-loop simulations with the sinusoidal noises are shown in Figs. 3.8 and 3.9. One can notice that the simulations with the non-Gaussian noises are similar to the Gaussian case. The $\mathcal{H}_\infty$ filter's response converges faster than the Kalman filter's response.

The estimation errors for the SMC-based Kalman and $\mathcal{H}_\infty$ filters for non-Gaussian noises are shown in Fig. 3.10. The RMSE plots are shown in Fig. 3.11, where the SMC based on $\mathcal{H}_\infty$ filter shows the better performance in the presence of the non-Gaussian noises. The small offsets are due to the non-zero bias of the noises, which were intentionally added to verify the effectiveness of the $\mathcal{H}_\infty$ filter in the presence of non-zero-mean noises. The $\infty - norms$ of state estimation errors over the simulation time for SMC-based Kalman filter's four states are 1.2004, 1.3677, 2.9674 and 1.4793, respectively and for the SMC-based $\mathcal{H}_\infty$ filter are 0.7185, 0.7254, 1.6648 and 0.8921, respectively.

3.6 Summary

In this chapter, the basic concepts and algorithms for the Kalman and $\mathcal{H}_\infty$ filters, and their application in the control theory have been presented. A combined sliding mode control and $\mathcal{H}_\infty$ filter scheme for practical systems are proposed. The proposed scheme is implemented on an illustration example of a full, nonlinear quadruple-tank system. The first two states of the quadruple-tank are assumed to be available and the remaining two states are estimated using the filters; the estimated states are then used for the sliding mode control design. The efficacy of the proposed approach for the quadruple-tank is verified by extensive simulations. The Kalman and $\mathcal{H}_\infty$ filters based sliding mode control were compared and it was found that the $\mathcal{H}_\infty$ filter based sliding mode control outperformed the Kalman filter based sliding mode control. It has also been shown that the proposed scheme not only works for Gaussian and non-Gaussian noises, but also works for non-zero-mean noises. This chapter mainly explores the linear state estimation methods and their applicability to the control theory; the next chapters will focus on nonlinear state estimation methods and their applications.

References

Banavar RN, Speyer JL (1991) A linear-quadratic game approach to estimation and smoothing. In: 1991 American control conference, pp 2818–2822. IEEE

Bandyopadhyay B, Fulwani D (2009) High-performance tracking controller for discrete plant using nonlinear sliding surface. IEEE Trans Ind Electron 56(9):3628–3637

Bandyopadhyay B, Deepak F, Kim K-S (2009) Sliding mode control using novel sliding surfaces, vol 392. Springer, Berlin

Basar T (1991) Optimum performance levels for minimax filters, predictors and smoothers. Syst Control Lett 16(5):309–317

Biswas PP, Srivastava R, Ray S, Samanta AN (2009) Sliding mode control of quadruple tank process. Mechatronics 19(4):548–561

Chandra KPB (2013) Nonlinear state estimation algorithms and their applications. PhD thesis, University of Leicester

Cheng G, Peng K (2007) Robust composite nonlinear feedback control with application to a servo positioning system. IEEE Trans Ind Electron 54(2):1132–1140

Csorba M (1998) Simultaneous localisation and map building. PhD thesis, University of Oxford

Durrant-Whyte H (2001) Introduction to estimation and the kalman filter. Aust Cent Field Robot 28(3):65–94

Einicke GA, White LB (1999) Robust extended kalman filtering. IEEE Trans Signal Process 47(9):2596–2599

Eustice RM, Singh H, Leonard JJ, Walter MR (2006) Visually mapping the RMS Titanic: conservative covariance estimates for slam information filters. Int J Robot Res 25(12):1223–1242

Grimble MJ, El Sayed A (1990) Solution of the h_∞ optimal linear filtering problem for discrete-time systems. IEEE Trans Acoust Speech Signal Process 38(7):1092–1104

Imtiaz SA, Roy K, Huang B, Shah SL, Jampana P (2006) Estimation of states of nonlinear systems using a particle filter. In: 2006 IEEE international conference on industrial technology, ICIT 2006, pp 2432–2437. IEEE

Johansson KH (2000) The quadruple-tank process: a multivariable laboratory process with an adjustable zero. IEEE Trans Control Syst Technol 8(3):456–465

Johnsen JK, Allöwer F (2007) Interconnection and damping assignment passivity-based control of a four-tank system. In: Lagrangian and hamiltonian methods for nonlinear control 2006, pp 111–122. Springer, Berlin

Julier SJ, Uhlmann JK (2004) Unscented filtering and nonlinear estimation. Proc IEEE 92(3):401–422

Kailath T (1968) An innovations approach to least-squares estimation-part i: linear filtering in additive white noise. IEEE Trans Autom Control 13(6):646–655

Kailath T, Sayed AH, Hassibi B (2000) Linear estimation, vol 1. Prentice Hall, Upper Saddle River

Kalman RE (1960) A new approach to linear filtering and prediction problems. J Basic Eng 82(1):35–45

Kolmogorov AN, Doyle WL, Selin I (1962) Interpolation and extrapolation of stationary random sequences. Bulletin of academic sciences, Mathematics series, USSR, vol 5, 1941. Translation by W. Doyle and J. Selin, RM-3090-PR, Rand Corporation, Santa Monica, California

Mercangöz M, Doyle FJ (2007) Distributed model predictive control of an experimental four-tank system. J Process Control 17(3):297–308

Mutambara AG (1998) Decentralized estimation and control for multisensor systems. CRC press, Boca Raton

Napgal K, Khargonekar P (1991) Filtering and smoothing in an h-infinity setting. IEEE Trans Autom Control 36:152–166

Okpanachi AV (2010) Developing advanced control strategies for a 4-tank laboratory process. Master's thesis report

Park P, Kailath T (1995) New square-root algorithms for kalman filtering. IEEE Trans Autom Control 40(5):895–899

Raff T, Huber S, Nagy ZK, Allgower F (2006) Nonlinear model predictive control of a four tank system: an experimental stability study. In: 2006 IEEE international conference on control applications, CCA'06, pp 237–242. IEEE

Said SH, M'Sahli F (2008) A set of observers design to a quadruple tank process. In: 2008 IEEE international conference on control applications, CCA 2008, pp 954–959. IEEE

Shaked U (1990) h_∞-minimum error state estimation of linear stationary processes. IEEE Trans Autom Control 35(5):554–558

Shen X-M, Deng L (1997) Game theory approach to discrete h_∞ filter design. IEEE Trans Signal Process 45(4):1092–1095

Simon D (2006) Optimal state estimation: Kalman, H infinity, and nonlinear approaches. Wiley, New York

Utkin V, Guldner J, Shi J (2009) Sliding mode control in electro-mechanical systems, vol 34. CRC press, Boca Raton

Vadigepalli R, Gatzke EP, Doyle FJ (2001) Robust control of a multivariable experimental four-tank system. Ind Eng Chem Res 40(8):1916–1927

Van Der Merwe R (2004) Sigma-point kalman filters for probabilistic inference in dynamic state-space models

Wiener N (1949) Extrapolation, interpolation, and smoothing of stationary time series, vol 7. MIT press, Cambridge

Yaesh I, Shaked U (1992) Game theory approach to optimal linear state estimation and its relation to the minimum h_∞ norm estimation. IEEE Trans Autom Control 37(6):828–831

Yang F, Wang Z, Lauria S, Liu X (2009) Mobile robot localization using robust extended h8 filtering. Proc Inst Mech Eng Part I J Syst Control Eng 223(8):1067–1080

Chapter 4
Jacobian-Based Nonlinear State Estimation

4.1 Introductory Remarks

Nonlinearity is one of the most challenging issues in controllers and observers design. Most of the practical systems are inherently nonlinear (Slotine et al. 1991; Khalil 2002). On the other hand, most control schemes and observer/estimater design methods used widely are for linear systems and are therefore not applicable in practice because of the existence of nonlinearities. This chapter deals with the state estimation methods for nonlinear systems stemmed from, however, linear estimator design ideas. Namely, the extensions of Kalman filter, information filter and an $\mathcal{H}_\infty$ filter to nonlinear systems will be discussed in this chapter. These extensions are based on the first-order Taylor's series expansion of the nonlinear process and/or measurement dynamics. The idea is, in some key steps of estimation, to replace the nonlinear process and/or measurement models with their corresponding Jacobians. The three filters to be discussed in this chapter are therefore the extended Kalman filter (EKF), extended information filter (EIF) and extended $\mathcal{H}_\infty$ filter (E$\mathcal{H}_\infty$F).

4.2 Extended Kalman Filter

Consider the following discrete, nonlinear process and measurement models

$$\mathbf{x}_k = \mathbf{f}(\mathbf{x}_{k-1}, \mathbf{u}_{k-1}) + \mathbf{w}_{k-1} \tag{4.1}$$

$$\mathbf{z}_k = \mathbf{h}(\mathbf{x}_k, \mathbf{u}_k) + \mathbf{v}_k \tag{4.2}$$

where k is the time index, $\mathbf{x}_k \in \mathbb{R}^n$ is the state vector, $\mathbf{u}_k$ the input variable, $\mathbf{z}_k$ the measurement (measurable output), $\mathbf{w}_{k-1}$ the process noise and $\mathbf{v}_k$ the measurement noise. The noises $\mathbf{w}_{k-1}$ and $\mathbf{v}_k$ are assumed to be zero mean, Gaussian-distributed, random variables with covariances of $\mathbf{Q}_{k-1}$ and $\mathbf{R}_k$, respectively.

K. P. B. Chandra and D.-W. Gu, *Nonlinear Filtering*,
https://doi.org/10.1007/978-3-030-01797-2_4

Similar to the Kalman filter, the extended Kalman filter (EKF) is also a recursive process consisting of prediction and measurement update but it requires Jacobians of the process and measurement dynamics.

To deal with nonlinear functions (or so-called nonlinear mappings), a natural and logical idea is to linearise the nonlinear functions and then to consider the " approximated" versions, i.e. the linearised ones, of the nonlinear functions instead. In doing so, the approximation should be kept as close as possible to the original function, for obvious reasons. In the estimation of nonlinear systems states, therefore, it is not directly to linearise the functions $\mathbf{f}(\cdot,\cdot)$ and $\mathbf{h}(\cdot,\cdot)$ in (4.1) and (4.2), respectively, at chosen equilibrium points and then to apply linear state estimation algorithms. Instead, to employ the nonlinear functions as longer as possible in the estimation and only to use the approximations of nonlinear functions when absolutely necessary.

In EKF, prediction of the state vector and covariance matrix can be calculated from

$$\hat{\mathbf{x}}_{k|k-1} = \mathbf{f}(\hat{\mathbf{x}}_{k-1|k-1}, \mathbf{u}_{k-1}) \tag{4.3}$$

$$\mathbf{P}_{k|k-1} = \nabla\mathbf{f}_x \mathbf{P}_{k-1|k-1} \nabla\mathbf{f}_x^T + \mathbf{Q}_{k-1} \tag{4.4}$$

where, $\nabla\mathbf{f}_x$ is the Jacobian of function $\mathbf{f}$ with respect to the state variable $\mathbf{x}$, obtained from the Taylor expansion series of $\mathbf{f}$.

The updated state and covariance can therefore be obtained as

$$\hat{\mathbf{x}}_{k|k} = \hat{\mathbf{x}}_{k|k-1} + \mathbf{K}_k[\mathbf{z}_k - \mathbf{h}(\hat{\mathbf{x}}_{k|k-1}, u_k)] \tag{4.5}$$

$$\mathbf{P}_{k|k} = (\mathbf{I}_n - \mathbf{K}_k \nabla\mathbf{h}_x)\mathbf{P}_{k|k-1} \tag{4.6}$$

where $\nabla\mathbf{h}_x$ is the Jacobian of function $\mathbf{h}$ with respect to $\mathbf{x}$ and $\mathbf{I}_n$ denotes the identity matrix of dimension $n \times n$.

The Kalman gain matrix is then given by

$$\mathbf{K}_k = \mathbf{P}_{k|k-1} \nabla\mathbf{h}_x^T \left[\nabla\mathbf{h}_x \mathbf{P}_{k|k-1} \nabla\mathbf{h}_x^T + \mathbf{R}_k\right]^{-1}. \tag{4.7}$$

The Jacobians $\nabla\mathbf{f}_x$ and $\nabla\mathbf{h}_x$, are evaluated at $\hat{\mathbf{x}}_{k-1|k-1}$ and at $\hat{\mathbf{x}}_{k|k-1}$, respectively.

The EKF procedure can now be summarised in Algorithm 4.1.

In comparison to the standard Kalman filter algorithm in Algorithm 3.1, it can be seen that in the prediction of $\hat{\mathbf{x}}_{k|k-1}$ and updating of $\hat{\mathbf{x}}_{k|k}$, both the original nonlinear functions $\mathbf{f}$ and $\mathbf{h}$ are used. However, in the derivation of the predicted state estimation error covariance $\mathbf{P}_{k|k-1}$, the difference between the state function $\mathbf{f}$ evaluated at the unavailable state $\mathbf{x}_{k-1}$ and at the predicted state $\hat{\mathbf{x}}_{k-1|k-1}$ is approximated by the Jacobian of $\mathbf{f}$, with regard to the state variable $\mathbf{x}$, multiplied by the increment $(\mathbf{x}_{k-1} - \hat{\mathbf{x}}_{k-1|k-1})$.

Similarly, in the updating of $\mathbf{P}_{k|k}$, the Jacobian $\nabla\mathbf{h}_x$ is employed in evaluating the difference between the "clean" measurement $\mathbf{h}$ at $\mathbf{x}_k$ and $\mathbf{h}$ at the estimated state $\hat{\mathbf{x}}_{k|k-1}$. Consequently $\nabla\mathbf{h}_x$ appears in the updating of the Kalman gain matrix as well.

4.3 Extended Information Filter

This section presents the idea of the extended information filter (EIF). EIF is an algebraic equivalent of EKF, in which the parameters of interest are *information states* and the inverse of the covariance matrix (*information matrix*) rather than the states and covariance matrix. EIF can be represented by a recursive process of prediction and measurement updates. The EIF equations are summarised below.

Consider the discrete nonlinear process and measurement models as

$$\mathbf{x}_k = \mathbf{f}(\mathbf{x}_{k-1}, \mathbf{u}_{k-1}) + \mathbf{w}_{k-1} \tag{4.8}$$

$$\mathbf{z}_k = \mathbf{h}(\mathbf{x}_k, \mathbf{u}_k) + \mathbf{v}_k \tag{4.9}$$

where k is the time index, $\mathbf{x}_k$ the state vector, $\mathbf{u}_k$ the control input, $\mathbf{z}_k$ the measurement. $\mathbf{w}_{k-1}$ and $\mathbf{v}_k$ are the process and measurement noises, respectively. These noises are assumed to be zero mean, Gaussian-distributed, random variables with covariances of $\mathbf{Q}_{k-1}$ and $\mathbf{R}_k$.

The conventional filter deals with the estimation of state vector, $\hat{\mathbf{x}}$ along with the corresponding variance matrix, $\mathbf{P}$. Whereas, the information filter deals with the information state, $\mathbf{y}$, and the corresponding information matrix (the inverse of the covariance matrix), $\mathbf{Y}$. The predicted information state vector, $\hat{\mathbf{y}}_{k|k-1}$, and the predicted information matrix, $\mathbf{Y}_{k|k-1}$, are given as

---

Algorithm 4.1 Extended Kalman filter

Initialise the state vector, $\hat{\mathbf{x}}_{0|0}$, and the covariance matrix, $\mathbf{P}_{0|0}$ (set $k = 1$).

Prediction

1: The predicted state and covariance matrix are

$$\hat{\mathbf{x}}_{k|k-1} = \mathbf{f}(\hat{\mathbf{x}}_{k-1|k-1}, \mathbf{u}_{k-1})$$
$$\mathbf{P}_{k|k-1} = \nabla\mathbf{f}_x \mathbf{P}_{k-1|k-1} \nabla\mathbf{f}_x^T + \mathbf{Q}_{k-1}.$$

Measurement update

1: The updated state and covariance can be obtained as

$$\hat{\mathbf{x}}_{k|k} = \hat{\mathbf{x}}_{k|k-1} + \mathbf{K}_k \left[\mathbf{z}_k - \mathbf{h}(\hat{\mathbf{x}}_{k|k-1}, u_k)\right]$$
$$\mathbf{P}_{k|k} = (\mathbf{I}_n - \mathbf{K}_k \nabla\mathbf{h}_x) \mathbf{P}_{k|k-1}$$

where the Kalman gain is

$$\mathbf{K}_k = \mathbf{P}_{k|k-1} \nabla\mathbf{h}_x^T \left(\nabla\mathbf{h}_x \mathbf{P}_{k|k-1} \nabla\mathbf{h}_x^T + \mathbf{R}_k\right)^{-1}.$$

$$\hat{\mathbf{y}}_{k|k-1} = \mathbf{Y}_{k|k-1}\hat{\mathbf{x}}_{k|k-1} \tag{4.10}$$

$$\mathbf{Y}_{k|k-1} = \mathbf{P}_{k|k-1}^{-1}$$

$$= \left[\nabla\mathbf{f}_x \mathbf{Y}_{k-1|k-1}^{-1} \nabla\mathbf{f}_x^T + \mathbf{Q}_{k-1}\right]^{-1} \tag{4.11}$$

where $\mathbf{P}_{k|k-1}$ is the predicted covariance matrix and

$$\hat{\mathbf{x}}_{k|k-1} = \mathbf{f}(\hat{\mathbf{x}}_{k-1|k-1}, \mathbf{u}_{k-1}). \tag{4.12}$$

The updated information state vector, $\hat{\mathbf{y}}_{k|k}$, and the updated information matrix, $\mathbf{Y}_{k|k}$, are

$$\hat{\mathbf{y}}_{k|k} = \hat{\mathbf{y}}_{k|k-1} + \mathbf{i}_k \tag{4.13}$$

$$\mathbf{Y}_{k|k} = \mathbf{Y}_{k|k-1} + \mathbf{I}_k. \tag{4.14}$$

The information state contribution, $\mathbf{i}_k$, and its associated information matrix, $\mathbf{I}_k$, are

$$\mathbf{i}_k = \nabla\mathbf{h}_x^T \mathbf{R}_k^{-1}\left[\nu_k + \nabla\mathbf{h}_x \hat{\mathbf{x}}_{k|k-1}\right] \tag{4.15}$$

$$\mathbf{I}_k = \nabla\mathbf{h}_x^T \mathbf{R}_k^{-1} \nabla\mathbf{h}_x \tag{4.16}$$

where the measurement residual, ν_k, is

$$\nu_k = \mathbf{z}_k - \mathbf{h}(\hat{\mathbf{x}}_{k|k-1}, \mathbf{u}_k) \tag{4.17}$$

and $\nabla\mathbf{f}_x$, and $\nabla\mathbf{h}_x$ are the Jacobians of $\mathbf{f}$ and $\mathbf{h}$ evaluated at the latest available state (Jacobians for prediction and update equations are evaluated at $\hat{\mathbf{x}}_{k-1|k-1}$ and $\hat{\mathbf{x}}_{k|k-1}$, respectively).

One of the key advantages of the information filter over the Kalman filter is during the update stage, where the updated information state and information matrix can be obtained by simply adding the associated information contributions to the predicted information state and information matrix. Detailed derivation can be found in Mutambara (1998).

For the extended information filter, recovery of state and covariance matrices are required at different stages and is still an active research area (Kaess and Dellaert 2009; Walter et al. 2007; Eustice et al. 2006; Huang et al. 2008). The state vector and covariance matrix can be recovered by using left division[1] Eustice et al. (2006)

$$\hat{\mathbf{x}}_{k|k} = \mathbf{Y}_{k|k} \backslash \hat{\mathbf{y}}_{k|k} \tag{4.18}$$

$$\mathbf{P}_{k|k} = \mathbf{Y}_{k|k} \backslash \mathbf{I}_n \tag{4.19}$$

where $\mathbf{I}_n$ is the state vector sized identity matrix.

[1] $\mathbf{x} = A\backslash B$ solves the least square solution for $A\mathbf{x} = B$ such that $\|A\mathbf{x} - b\|$ is minimal.

Algorithm 4.2 Extended Information Filter

Initialise the information vector, $\hat{\mathbf{y}}_{0|0}$, and the information matrix, $\mathbf{Y}_{0|0}$ (set $k = 1$).

Prediction

1: The predicted information vector and information matrix are

$$\hat{\mathbf{y}}_{k|k-1} = \mathbf{Y}_{k|k-1}\hat{\mathbf{x}}_{k|k-1}$$
$$\mathbf{Y}_{k|k-1} = \mathbf{P}^{-1}_{k|k-1}$$
$$= \left[\nabla\mathbf{f}_x \mathbf{Y}^{-1}_{k-1|k-1} \nabla\mathbf{f}_x^T + \mathbf{Q}_{k-1}\right]^{-1}$$

where

$$\hat{\mathbf{x}}_{k|k-1} = \mathbf{f}(\mathbf{Y}_{k-1|k-1} \backslash \mathbf{y}_{k-1|k-1}, \mathbf{u}_{k-1}).$$

Measurement update

1: The updated information vector and information matrix are

$$\hat{\mathbf{y}}_{k|k} = \hat{\mathbf{y}}_{k|k-1} + \mathbf{i}_k$$
$$\mathbf{Y}_{k|k} = \mathbf{Y}_{k|k-1} + \mathbf{I}_k.$$

where

$$\mathbf{i}_k = \nabla\mathbf{h}_x^T \mathbf{R}_k^{-1}\left[\nu_k + \nabla\mathbf{h}_x \hat{\mathbf{x}}_{k|k-1}\right]$$
$$\mathbf{I}_k = \nabla\mathbf{h}_x^T \mathbf{R}_k^{-1} \nabla\mathbf{h}_x$$

and

$$\nu_k = \mathbf{z}_k - \mathbf{h}(\hat{\mathbf{x}}_{k|k-1}, \mathbf{u}_k)$$

Initialisation in the information filter is easier than that in the Kalman filter and the update stage of information filter is computationally simpler than that of the Kalman filter. EIF can be shown to be more efficient than the EKF. But some of the drawbacks inherent in the EKF still affect the EIF. These include the nontrivial nature of the derivations of the Jacobian matrices (and computation) and linearisation instability (Mutambara 1998).

Since the EIF is an alternate form of EKF, the simulation results for the EIF is not too different to the one given in Figs. 4.1 and 4.2. However, the key advantage of the EIF is its usage in multi-sensor estimation; that will be further demonstrated in later chapters.

4.4 Extended $\mathcal{H}_\infty$ Filter

This section presents the extended $\mathcal{H}_\infty$ filter (E$\mathcal{H}_\infty$F). As discussed previously about the linear systems, the stochastic properties of the noises affecting the model and output dynamics may not be exactly known. Hence, the Kalman filters, either the standard or extended versions, will not be applicable in such cases. Therefore, $\mathcal{H}_\infty$ filters could be a tool worth trying. The $\mathcal{H}_\infty$ filters do not need to assume that the process and measurement noises be Gaussian. Furthermore, the $\mathcal{H}_\infty$ filters could tolerate a certain degree of dynamic uncertainties in the process and measurement models. Similar to the linear case, the extended $\mathcal{H}_\infty$ filter (E$\mathcal{H}_\infty$F) minimises the expected value of the variance of the estimation error and uses the Jacobian to approximate the differences of the process (measurement) function evaluated at the unavailable state variable $\mathbf{x}_{k-1}$ ($\mathbf{x}_k$) and at the predicted state $\hat{\mathbf{x}}_{k-1|k-1}$ ($\hat{\mathbf{x}}_{k|k-1}$).

Consider the discrete-time process and observation models written as

$$\mathbf{x}_k = \mathbf{f}[\mathbf{x}_{k-1}, \mathbf{u}_{k-1}] + \mathbf{w}_{k-1} \tag{4.20}$$

$$\mathbf{z}_k = \mathbf{h}[\mathbf{x}_k, \mathbf{u}_k] + \mathbf{v}_k \tag{4.21}$$

where k is the current time index, $\mathbf{x}_k \in \mathcal{R}^n$ is a state vector, $\mathbf{u}_k \in \mathcal{R}^q$ a control input, $\mathbf{z}_k \in \mathcal{R}^p$ a measurement vector, and $\mathbf{w}_{k-1}$ and $\mathbf{v}_k$ are the process and observation noises. The noise terms $\mathbf{w}_k$ and $\mathbf{v}_k$ may be random with possibly unknown statistics, or even they may be deterministic. And, they may have a non-zero mean. Instead of directly estimating the state one could estimate a linear combination of states (Einicke and White 1999; Yang et al. 2009).

In the game theory approach to $\mathcal{H}_\infty$ filtering (Banavar and Speyer 1991; Yang et al. 2009, the performance measure is given by

$$\mathbf{J}_\infty = \frac{\sum_{k=1}^{N} \|\mathbf{n}_k - \widehat{\mathbf{n}}_k\|^2_{\mathbf{M}_k}}{\|\mathbf{x}_0 - \hat{\mathbf{x}}_0\|^2_{\mathbf{P}_0^{-1}} + \sum_{k=1}^{N}(\|\mathbf{w}_k\|^2_{\mathbf{Q}_k^{-1}} + \|\mathbf{v}_k\|^2_{\mathbf{R}_k^{-1}})} \tag{4.22}$$

where $\mathbf{P}_0, \mathbf{Q}_k, \mathbf{R}_k$, and $\mathbf{M}_k$ are symmetric, positive definite, weighting matrices chosen by the user based on the problem at hand. The norm notation used in this section is $\|e\|^2_{S_k} = e^T S_k e$. This is the same performance measure which was discussed in Sect. 3.4 of Chap. 3 for linear systems. A linear $\mathcal{H}_\infty$ filter can be easily extended to nonlinear systems by replacing the linear state and measurement matrices with their Jacobians, and by slightly modifying the predicted state equation. In this book, we introduce the E$\mathcal{H}_\infty$F algorithm given in Yang et al. (2009).

The predicted state vector and auxiliary matrix are

$$\widehat{\mathbf{x}}_{k|k-1} = \mathbf{f}(\widehat{\mathbf{x}}_{k-1|k-1}, \mathbf{u}_{k-1})$$
$$\mathbf{P}_{k|k-1} = \nabla\mathbf{f}_x \mathbf{P}_{k-1|k-1} \nabla\mathbf{f}_x^T + \mathbf{Q}_k$$

and the inverse of the updated auxiliary matrix can be obtained as

Algorithm 4.3 Extended $\mathcal{H}_\infty$ filter

Initialise the state vector, $\hat{\mathbf{x}}_{0|0}$, and the auxiliary matrix, $\mathbf{P}_{0|0}$ (set $k = 1$).

Prediction

1: The predicted state and auxiliary matrix are

$$\hat{\mathbf{x}}_{k|k-1} = \mathbf{f}(\hat{\mathbf{x}}_{k-1|k-1}, \mathbf{u}_{k-1})$$
$$\mathbf{P}_{k|k-1} = \nabla\mathbf{f}_x \mathbf{P}_{k-1|k-1} \nabla\mathbf{f}_x^T + \mathbf{Q}_{k-1}.$$

Measurement update

1: The updated state and auxiliary matrix are

$$\hat{\mathbf{x}}_{k|k} = \hat{\mathbf{x}}_{k|k-1} + \mathbf{K}_{\infty,k}\left[\mathbf{z}_k - \mathbf{h}(\hat{\mathbf{x}}_{k|k-1})\right]$$
$$\mathbf{P}_{k|k} = (\mathbf{I}_n - \mathbf{K}_k \nabla\mathbf{h}_x)\,\mathbf{P}_{k|k-1} - \gamma^{-2}\mathbf{I}_n$$

where the gain

$$\mathbf{K}_{\infty,k} = \mathbf{P}_{k|k-1}\nabla\mathbf{h}_x^T\left(\nabla\mathbf{h}_x\mathbf{P}_{k|k-1}\nabla\mathbf{h}_x^T + \mathbf{R}_k\right)^{-1}.$$

$$\mathbf{P}_{k|k}^{-1} = \mathbf{P}_{k|k-1}^{-1} + \nabla\mathbf{h}_x^T \mathbf{R}_k^{-1} \nabla\mathbf{h}_x - \gamma^{-2}\mathbf{I}_n$$

where $\mathbf{I}_n$ denotes the identity matrix of dimension $n \times n$.
The updated state is

$$\widehat{\mathbf{x}}_{k|k} = \widehat{\mathbf{x}}_{k|k-1} + \mathbf{K}_\infty[\mathbf{z}_k - \mathbf{h}(\widehat{\mathbf{x}}_{k|k-1})]$$

where

$$\mathbf{K}_\infty = \mathbf{P}_{k|k-1}\nabla\mathbf{h}_x^T[\nabla\mathbf{h}_x\mathbf{P}_{k|k-1}\nabla\mathbf{h}_x^T + \mathbf{R}_k]^{-1}$$

The Jacobians of $\mathbf{f}$ and $\mathbf{h}$, $\nabla\mathbf{f}_x$ and $\nabla\mathbf{h}_x$, are evaluated at $\widehat{\mathbf{x}}_{k-1|k-1}$ and at $\widehat{\mathbf{x}}_{k|k-1}$, respectively.

The extended $\mathcal{H}_\infty$ filter algorithm is summarised below.

4.5 A DC Motor Case Study

Example 4.1 Consider the direct current (DC) motor dynamics below Mehta and Chiasson (1998), a linearised model of which was used in Chap. 2,

Fig. 4.1 Actual and estimated velocity using EKF

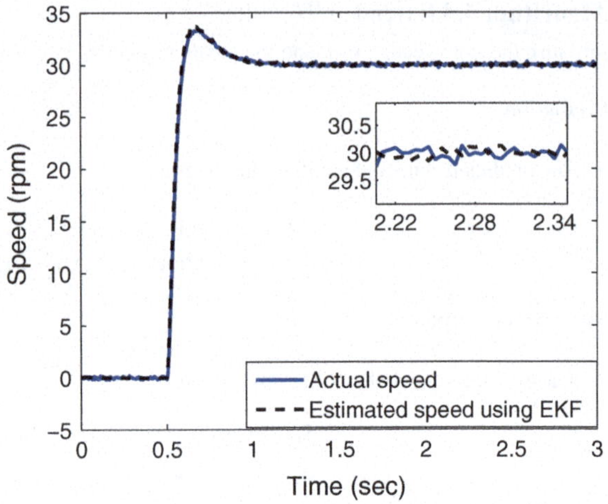

$$\frac{di}{dt} = \frac{1}{L}\left(-Ri - K_m L_f i\omega + V\right) + w_i \tag{4.23}$$

$$\frac{d\omega}{dt} = \frac{1}{J}\left(K_m L_f i^2 - D\omega - T_L\right) + w_\omega \tag{4.24}$$

where, i and w, current and velocity, are the states of the motor. The input voltage is represented by V, and T_L is the load torque. w_i and w_ω represents the Gaussian process noise. The remaining parameters of the motor are $R = 7.2\ \Omega$, $K_m L_f = 0.1236\,\text{Nm/WbA}$, $D = 0.0004\,\text{Hm/rad/s}$ and $J = 0.0007046\,\text{kgm}^2$. It is assumed that only the noisy current measurement is available, such that the output of the DC motor is $y = i + v_i$, where v_i is the Gaussian sensor noise. The covariance matrices for process and measurement models are $Q =$ and $R =$. The discrete-time model of DC motor in (4.23) and (4.24) is obtained by using Euler's method with sampling time $dt = 0.001\ s$. The required voltage V to maintain the motor at a desired velocity is generated by a PID controller, where the actual velocity is compared with the reference velocity and then based on the error the PID controller provides correspondingly voltage to the motor. In this example it is assumed that the velocity measurement is unavailable. An EKF is therefore used to estimate the velocity. Consequently, instead of actual velocity, an estimated velocity is used for the PID control. The estimated states and the actual states are shown in the Fig. 4.1 and the error between the actual velocity and estimated one is shown in Fig. 4.2. It can be seen that the designed EKF can estimate the speed with a reasonable accuracy.

The E$\mathcal{H}_\infty$F approach can now be applied to estimate the velocity of the DC motor. The attenuation parameter in Algorithm 4.3 is chosen as $\gamma = 0.01$, the remaining parameters are the same as those defined earlier with EKF approach. The estimated states and the actual states are shown in Fig. 4.3 and the error between the actual velocity and estimated velocity is shown in Fig. 4.4. It can be seen that the designed E$\mathcal{H}_\infty$F can estimate the velocity with a reasonable accuracy. This is mainly because

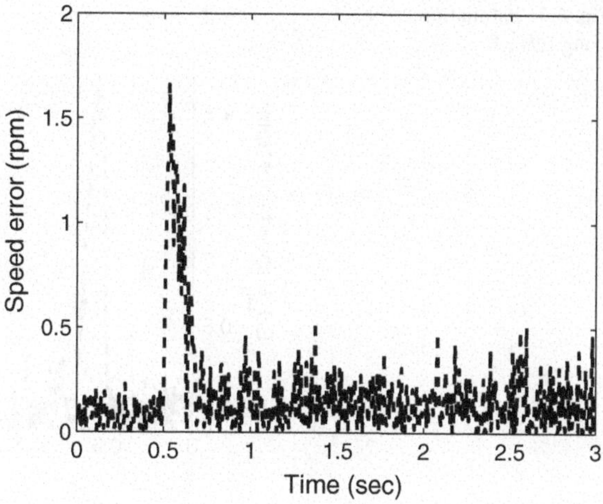

Fig. 4.2 Estimation error of EKF

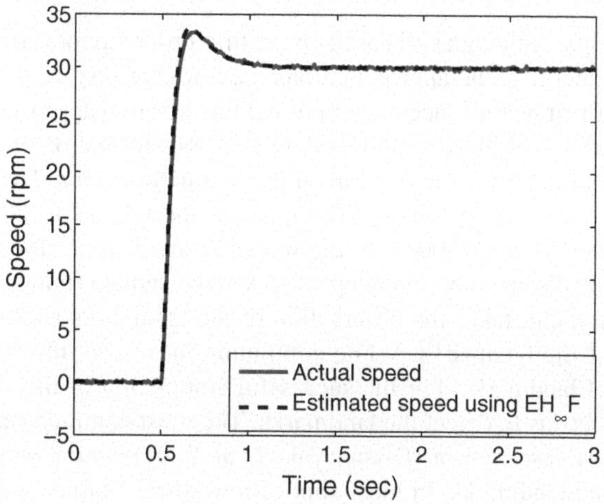

Fig. 4.3 Actual and estimated velocity using $E\mathcal{H}_\infty F$

the parameters of the DC motor model is assumed to be perfectly known. Of course, when the attenuation parameter γ is set to zero, then $E\mathcal{H}_\infty F$ and EKF are the same.

4.6 Nonlinear Transformation and the Effects of Linearisation

As mentioned earlier, direct linearisation of nonlinear system dynamic and/or of measurement functions is not recommendable. It is not just because such linearisation would be valid only around certain equilibrium points but also it could lead to low quality of estimation. The EKF described in Sect. 4.2 tries to avoid such shortcom-

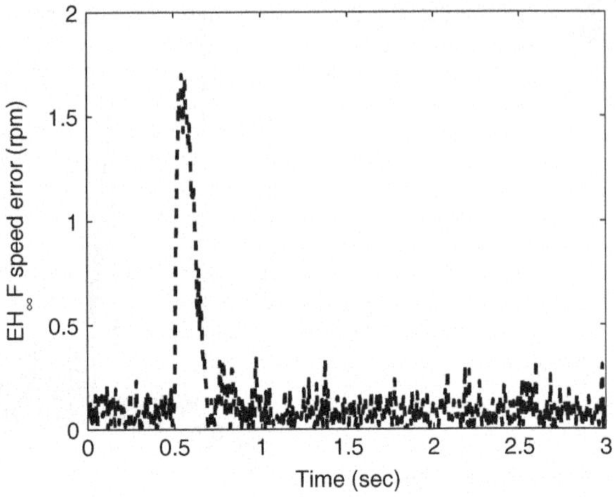

Fig. 4.4 Estimation error using $E\mathcal{H}_\infty F$

ings. Although still based on the first-order Taylor series approximation of nonlinear functions, the approximations are however just used in the iterative estimation. To illustrate the inaccuracy of direct linearisation approach, this section investigates the effects of linearisation on nonlinear transformation of polar to cartesian coordinates and analyses the corresponding estimation error. The similar discussion has been considered in Simon (2006), Julier and Uhlmann (2004), Julier et al. (2000), Van Der Merwe (2004). In the mapping application, for example in the simultaneous localisation and mapping (SLAM) the vehicle (maybe a robot or an uninhabited air vehicle) takes the observation of the landmarks and outputs the range and bearing of the landmarks. While continuing in motion, the vehicle builds a complete map of landmarks. For the successful completion of this mapping task, one of the key steps is to detect the landmarks. The most common sensor used in robotic mapping is a laser sensor (Dissanayake et al. 2001), which outputs the range and bearing of the landmarks. In most applications, these outputs will have to be converted to the cartesian coordinates for further analysis and control design. One may see Chandra (2013) for more details.

Consider the polar to cartesian nonlinear transformation given by

$$\begin{bmatrix} x \\ y \end{bmatrix} = \mathbf{h}(\mathbf{x}) = \begin{bmatrix} r\cos\theta \\ r\sin\theta \end{bmatrix} \tag{4.25}$$

where $\mathbf{x}$, consisting of range and bearing is $\mathbf{x} = [r \quad \theta]^T$, and $[x \quad y]^T$ are the cartesian coordinates of the target. It is assumed that r and θ are two independent variables with means $\bar{r}$ and $\bar{\theta}$, and the corresponding standard deviations are σ_r and σ_θ, respectively.

The range and bearings in the polar coordinates frame can be further written as

$$r = \bar{r} + r_e \tag{4.26}$$

$$\theta = \bar{\theta} + \theta_e \tag{4.27}$$

with $\bar{r} = 1$ and $\bar{\theta} = \frac{\pi}{2}$. r_e and θ_e are the corresponding zero-mean deviations from their means. It is assumed that r_e and θ_e are uniformly distributed[2] between $\pm r_m$ and $\pm\theta_m$, respectively. A similar scenario has been considered in Simon (2006).

The means of x and y, $\bar{x}$ and $\bar{y}$, can be obtained by taking the expectations of x and y as given below

$$\begin{aligned}
\bar{x} &= \mathbb{E}(r\cos\theta) \\
&= \mathbb{E}[(\bar{r} + r_e)(\cos(\bar{\theta} + \theta_e))] \\
&= \mathbb{E}[(\bar{r} + r_e)(\cos\bar{\theta}\cos\theta_e - \sin\bar{\theta}\sin\theta_e)] \\
&= \mathbb{E}[-\bar{r}\sin\bar{\theta}\sin\theta_e - r_e\sin\bar{\theta}\sin\theta_e] \\
&= \mathbb{E}[-\sin\theta_e] \qquad (\because\ r_e \text{ and } \theta_e \text{ are independent}) \\
&= \frac{1}{2\theta_m}[\cos\theta_e]_{-\theta_m}^{\theta_m} \qquad (\text{from } 4.28) \\
&= 0
\end{aligned} \tag{4.29}$$

and

$$\begin{aligned}
\bar{y} &= \mathbb{E}(r\sin\theta) \\
&= \mathbb{E}[(\bar{r} + r_e)\sin(\bar{\theta} + \theta_e)] \\
&= \mathbb{E}[\bar{r}\sin\bar{\theta}\cos\theta_e + r_e\sin\bar{\theta}\cos\theta_e + \bar{r}\cos\bar{\theta}\sin\theta_e + r_e\cos\bar{\theta}\sin\theta_e] \\
&= \mathbb{E}(\bar{r}\sin\bar{\theta}\cos\theta_e) \\
&= \mathbb{E}(\cos\theta_e) \\
&= \frac{1}{2\theta_m}[\sin\theta_e]_{-\theta_m}^{\theta_m} \qquad (\text{from } 4.28) \\
&= \frac{\sin\theta_m}{\theta_m} < 1.
\end{aligned} \tag{4.30}$$

Similarly, the covariance can be obtained as Simon (2006)

$$\begin{aligned}
\mathbf{P}_{x,y} &= \mathbb{E}\left[\begin{bmatrix}(x - \bar{x}) \\ (y - \bar{y})\end{bmatrix}\begin{bmatrix}(x - \bar{x}) \\ (y - \bar{y})\end{bmatrix}^T\right] \\
&= \mathbb{E}\left[\begin{bmatrix} r\cos\theta \\ r\sin\theta - \frac{\sin\theta_m}{\theta_m}\end{bmatrix}\begin{bmatrix} r\cos\theta \\ r\sin\theta - \frac{\sin\theta_m}{\theta_m}\end{bmatrix}^T\right]
\end{aligned}$$

---

[2] If a variable x is uniformly distributed between a and b, $\mathbb{U}(a, b)$, then the nth moment of x is

$$\mathbb{E}(x^n) = \frac{1}{b - a}\int_a^b x^n dx. \tag{4.28}$$

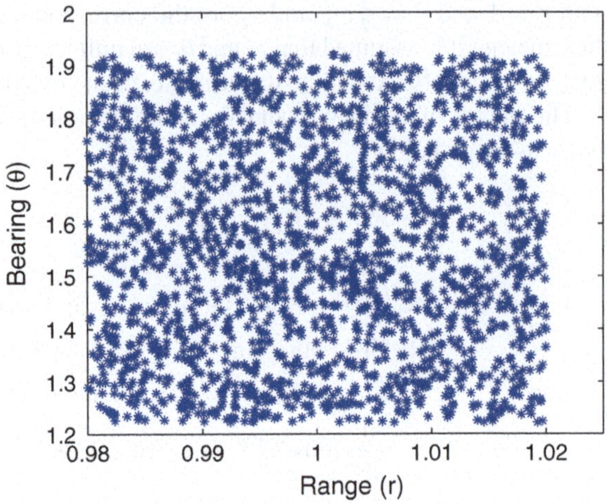

Fig. 4.5 2000 random points (∗) are generated with range and bearings, which are uniformly distributed between ±0.02 and ±20°

$$= \begin{bmatrix} \frac{1}{2}(1+\sigma_r^2)\left(1-\frac{\sin 2\theta_m}{2\theta_m}\right) & 0 \\ 0 & \frac{1}{2}(1+\sigma_r^2)\left(1+\frac{\sin 2\theta_m}{2\theta_m}\right)-\frac{\sin^2\theta_m}{\theta_m^2} \end{bmatrix} \tag{4.31}$$

Simulations have been performed to see the true mean and covariance ellipse of the nonlinear polar to cartesian coordinate transformation. In the rest of this section, the means of x and y given in Eqs. (4.29) and (4.30) are called as the true mean and the ellipse formed by the first and fourth elements of the $\mathbf{P}_{x,y}$ given in Eq. (4.31) is called as the true ellipse. 2000 measurement samples were generated by taking the true range and bearing values of the target location and adding a zero-mean r_e and θ_e, which are uniformly distributed between ± 0.02 and $\pm 20°$. The corresponding plot is shown in Fig. 4.5. The range is varying from $\bar{r} \pm r_m$ i.e. (1 ± 0.02) and the bearing is varying from $\bar{\theta} \pm \theta_m$ i.e. $\left(\frac{\pi}{2} \pm 0.3491\text{rad}\right)$. These random points are then processed through the nonlinear polar to cartesian coordinate transformation and are shown in Fig. 4.6. The nonlinear mean and the standard deviation ellipse are also shown in Fig. 4.6.

From Eqs. (4.29) and (4.30), the mean of x is 0 and for the y is less than 1; the same can be seen in Fig. 4.6, where the means of x and y are 0 and 0.9798 (which is less than 1), respectively.

Polar to Cartesian Coordinates Transformation: First-Order Linearisation

Now, the polar to cartesian coordinates transformation using first-order linearisation is to be investigated. The mean of Eq. (4.25) can be obtained by taking the expected values on both sides and can be written as

$$\mathbb{E}\begin{bmatrix} x \\ y \end{bmatrix} = \mathbb{E}\begin{bmatrix} r\cos\theta \\ r\sin\theta \end{bmatrix} \tag{4.32}$$

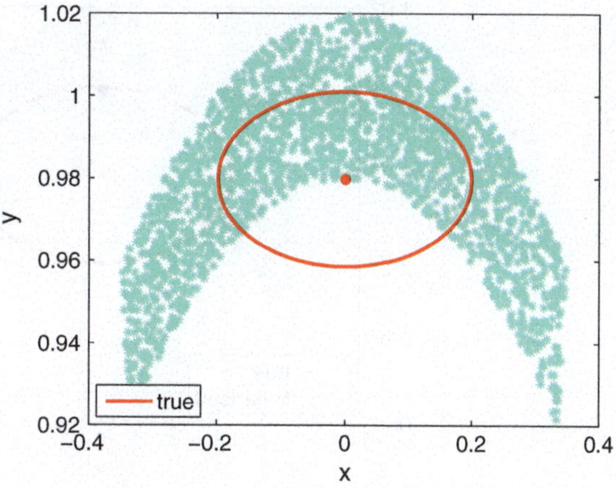

Fig. 4.6 2000 random measurements are generated with range and bearings, which are uniformly distributed between ±0.02 and ±20°. These random points are then processed through the nonlinear polar to cartesian coordinate transformation and are shown as ∗. The true mean and the uncertainty ellipse are represented by • and solid line, respectively

To analyse the effects of linearisation, the nonlinear terms in Eq. (4.32) are expanded using Taylor's series (the second and higher order derivative terms are neglected).

$$\bar{\mathbf{h}}(\mathbf{x}) \simeq \mathbb{E}\left[\begin{pmatrix}\bar{x}\\ \bar{y}\end{pmatrix} + \nabla_S\Big|_{\bar{\mathbf{x}}}\begin{pmatrix}x-\bar{x}\\ y-\bar{y}\end{pmatrix}\right] \tag{4.33}$$

$$= \begin{bmatrix}\bar{x}\\ \bar{y}\end{bmatrix} + \nabla_S\Big|_{\bar{r},\bar{\theta}}\mathbb{E}\begin{bmatrix}x-\bar{x}\\ y-\bar{y}\end{bmatrix} \tag{4.34}$$

$$= \begin{bmatrix}\bar{r}\cos\bar{\theta}\\ \bar{r}\sin\bar{\theta}\end{bmatrix} \tag{4.35}$$

$$= \begin{bmatrix}0\\ 1\end{bmatrix} \tag{4.36}$$

where $\nabla_{\mathbf{s}}\Big|_{\bar{r},\bar{\theta}}$ is the Jacobian of $\mathbf{s}$ evaluated at $(\bar{r},\bar{\theta})$ and is given by

$$\nabla_{\mathbf{s}}\Big|_{\bar{r},\bar{\theta}} = \begin{bmatrix}\cos\theta & -r\sin\theta\\ \sin\theta & r\cos\theta\end{bmatrix}\Bigg|_{\bar{r},\bar{\theta}} \tag{4.37}$$

$$= \begin{bmatrix}0 & -1\\ 1 & 0\end{bmatrix} \tag{4.38}$$

The linearised covariance Simon (2006) of Eq. (4.25) is

$$\mathbf{P} = \nabla_{\mathbf{s}}\Big|_{\bar{r},\bar{\theta}}\mathbf{P}_{r,\theta}\nabla_s\Big|_{\bar{r},\bar{\theta}}^{T} \tag{4.39}$$

where

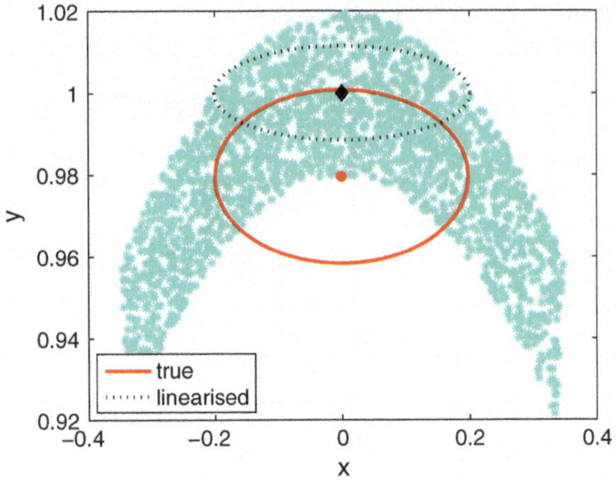

Fig. 4.7 2000 random measurements are generated with range and bearings, which are uniformly distributed between ± 0.02 and $\pm 20°$. These random points are then processed through the nonlinear polar to cartesian coordinate transformation and are shown as $*$. The true mean and the linearised mean are represented by • and ♦, and true and linearised uncertainty ellipses are represented by solid and dotted lines, respectively

$$\mathbf{P}_{r,\theta} = \mathbb{E}\left[\begin{pmatrix} r - \bar{r} \\ \theta - \bar{\theta} \end{pmatrix} \begin{pmatrix} r - \bar{r} \\ \theta - \bar{\theta} \end{pmatrix}^T\right] \tag{4.40}$$

$$= \begin{bmatrix} \sigma_r^2 & 0 \\ 0 & \sigma_\theta^2 \end{bmatrix} \tag{4.41}$$

and, σ_r and σ_θ are the standard deviations of r and θ, respectively.

The mean and the standard deviation ellipse in the linearisation case along with the true mean and true uncertainty ellipse are shown in Fig. 4.7. It can be seen that the linearised mean and standard deviation ellipse are not consistent with the true mean and true uncertainty ellipse. The true mean is located at (0, 0.9798), whereas the linearised mean is located at (0, 1).

4.7 Summary

Based on the introduction of mathematical preliminary and state estimation methods for linear control systems, this chapter discussed three approaches to estimate the states of nonlinear control systems. The three approaches, namely the extended Kalman filter (EKF), extended information filter (EIF) and extended $\mathcal{H}_\infty$ filter (E$\mathcal{H}_\infty$F), all use the Taylor series expansion to deal with nonlinear dynamics and/or nonlinear output (measurement) equations appeared in nonlinear control system models. The idea is natural and logical, and is generated from state estimation approaches for linear systems. Related algorithms and properties have been introduced in this chapter together with illustrative examples and simulations. Towards the end of the chapter, the accuracy property of the polar-cartesian coordinate transformation was considered. This is an interesting example which illustrates the potential drawbacks in using (direct) linearisation approach towards nonlinear functions. It therefore indicates that different approaches need to be explored in state estimation for nonlinear control systems other than EKF, EIF and $E\mathcal{H}_\infty F$ introduced in this chapter in which Taylor series expansion is used. Such developments are the theme of the book and will be discussed in the rest part of the book.

References

Banavar RN, Speyer JL (1991) A linear-quadratic game approach to estimation and smoothing. In: 1991 American control conference. IEEE, pp 2818–2822

Chandra KPB (2013) Nonlinear state estimation algorithms and their applications. PhD thesis, University of Leicester (2013)

Dissanayake MG, Newman P, Clark S, Durrant-Whyte HF, Csorba M (2001) A solution to the simultaneous localization and map building (slam) problem. IEEE Trans Robot Autom 17(3):229–241

Einicke GA, White LB (1999) Robust extended kalman filtering. IEEE Trans Signal Process 47(9):2596–2599

Eustice RM, Singh H, Leonard JJ, Walter MR (2006) Visually mapping the rms titanic: conservative covariance estimates for slam information filters. Int J Robot Res 25(12):1223–1242

Huang S, Wang Z, Dissanayake G (2008) Exact state and covariance sub-matrix recovery for submap based sparse eif slam algorithm. In: 2008 IEEE international conference on robotics and automation, ICRA 2008. IEEE, pp 1868–1873

Julier SJ, Uhlmann JK (2004) Unscented filtering and nonlinear estimation. Proc IEEE 92(3):401–422

Julier S, Uhlmann J, Durrant-Whyte HF (2000) A new method for the nonlinear transformation of means and covariances in filters and estimators. IEEE Trans Autom control 45(3):477–482

Kaess M, Dellaert F (2009) Covariance recovery from a square root information matrix for data association. Robot Auton Syst 57(12):1198–1210

Khalil HK (2002) Nonlinear Systems, 3rd edition, Prentice Hall, Upper Saddle River, New Jersey

Mehta S, Chiasson J (1998) Nonlinear control of a series dc motor: theory and experiment. IEEE Trans Ind Electron 45(1):134–141

Mutambara AG (1998) Decentralized estimation and control for multisensor systems. CRC press, Bocca Raton

Simon D (2006) Optimal state estimation: Kalman, H infinity, and nonlinear approaches. Wiley, New York

Slotine J-JE, Li W (1991) Applied nonlinear control, vol 199. Prentice hall, Englewood Cliffs

Van Der Merwe R (2004) Sigma-point kalman filters for probabilistic inference in dynamic state-space models

Walter MR, Eustice RM, Leonard JJ (2007) Exactly sparse extended information filters for feature-based slam. Int J Robot Res 26(4):335–359

Yang F, Wang Z, Lauria S, Liu X (2009) Mobile robot localization using robust extended h8 filtering. Proc Inst Mech Eng Part I: J Syst Control Eng 223(8):1067–1080

Chapter 5
Cubature Kalman Filter

5.1 Introductory Remarks

The cubature Kalman filter (*CKF*) is the closest approximation known so far to the Bayesian filter that could be designed in a nonlinear setting under the Gaussian assumption. Unlike the extended Kalman filter (EKF), CKF does not require evaluation of Jacobians during the estimation process, while in EKF the nonlinear functions are approximated by their Jacobians, the first-order Taylor's series approximation. The unscented Kalman filter (UKF) that is to be introduced in the last chapter is also a useful tool in nonlinear state estimation. However, UKF performance is completely dominated by the tuning parameters, α, β and κ. In comparison, CKF neither requires Jacobians nor depends on additional tuning parameters. CKF is therefore an appealing option for nonlinear state estimation to EKF or UKF (Arasaratnam and Haykin 2009). The idea, basic steps required in design and the properties of CKF are to be described in this chapter.

5.2 CKF Theory

Consider a nonlinear system with additive noise defined by process and measurement models in (4.1) and (4.2), listed below for convenience,

$$\mathbf{x}_k = \mathbf{f}(\mathbf{x}_{k-1}, \mathbf{u}_{k-1}) + \mathbf{w}_{k-1} \tag{5.1}$$

$$\mathbf{z}_k = \mathbf{h}(\mathbf{x}_k, \mathbf{u}_k) + \mathbf{v}_k \tag{5.2}$$

The key assumption of the cubature Kalman filter is that the predictive density $p(\mathbf{x}_k|D_{k-1})$, where $D_{k-1} = (\mathbf{u}_l, \mathbf{z}_l)_{l=1}^{k-1}$ denotes the history of input-measurement pairs up to $k-1$ inclusively, and the filter likelihood density $p(\mathbf{z}_k|D_k)$ are both Gaussian, which eventually leads to a Gaussian posterior density $p(\mathbf{x}_k|D_k)$. Under this assumption, the CKF solution reduces to how to compute their means and covariances more accurately.

K. P. B. Chandra and D.-W. Gu, *Nonlinear Filtering*,
https://doi.org/10.1007/978-3-030-01797-2_5

The CKF is a two-stage procedure comprising of prediction and update, the same as in most state estimators.

5.2.1 Prediction

At the prediction stage, the CKF computes the mean $\hat{\mathbf{x}}_{k|k-1}$ of the predicted states and the associated covariance $\mathbf{P}_{k|k-1}$ of the Gaussian predictive density numerically using cubature rules. The predicted mean can be written as

$$\hat{\mathbf{x}}_{k|k-1} = \mathbb{E}\left[\mathbf{f}(\mathbf{x}_{k-1}, \mathbf{u}_{k-1}) + \mathbf{w}_{k-1}|D_{k-1}\right] \tag{5.3}$$

Since $\mathbf{w}_{k-1}$ is assumed to be zero-mean and uncorrelated with the measurement sequence, it yields therefore

$$\begin{aligned}\hat{\mathbf{x}}_{k|k-1} &= \mathbb{E}\left[\mathbf{f}(\mathbf{x}_{k-1}, \mathbf{u}_{k-1})|D_{k-1}\right]\\ &= \int_{R^n} \mathbf{f}(\mathbf{x}_{k-1}, \mathbf{u}_{k-1})p(\mathbf{x}_{k-1}|D_{k-1})d\mathbf{x}_{k-1}\\ &= \int_{R^n} \mathbf{f}(\mathbf{x}_{k-1}, \mathbf{u}_{k-1})\mathcal{N}(\mathbf{x}_{k-1}; \hat{\mathbf{x}}_{k-1|k-1}, \mathbf{P}_{k-1|k-1})d\mathbf{x}_{k-1}.\end{aligned} \tag{5.4}$$

Similarly, the associated error covariance can be represented as

$$\begin{aligned}\mathbf{P}_{k|k-1} &= \mathbb{E}\left[(\mathbf{x}_k - \hat{\mathbf{x}}_{k|k-1})(\mathbf{x}_k - \hat{\mathbf{x}}_{k|k-1})^T|\mathbf{z}_{k-1}\right]\\ &= \int_{R^n} \mathbf{f}(\mathbf{x}_{k-1}, \mathbf{u}_{k-1})\mathbf{f}^T(\mathbf{x}_{k-1}, \mathbf{u}_{k-1})\mathcal{N}(\mathbf{x}_{k-1}; \hat{\mathbf{x}}_{k-1|k-1}, \mathbf{P}_{k-1|k-1})d\mathbf{x}_{k-1}\\ &\quad -\hat{\mathbf{x}}_{k|k-1}\hat{\mathbf{x}}_{k|k-1}^T + \mathbf{Q}_{k-1}.\end{aligned} \tag{5.5}$$

5.2.2 Measurement Update

The predicted measurement density can be represented by

$$p(\mathbf{z}_k|D_{k-1}) = \mathcal{N}(\mathbf{z}_k; \hat{\mathbf{z}}_{k|k-1}, \mathbf{P}_{zz,k|k-1}) \tag{5.6}$$

where the predicted measurement and associated covariance are given by

$$\hat{\mathbf{z}}_{k|k-1} = \int_{R^n} \mathbf{h}(\mathbf{x}_k, \mathbf{u}_k)\mathcal{N}(\mathbf{x}_k; \hat{\mathbf{x}}_{k|k-1}, \mathbf{P}_{k|k-1})d\mathbf{x}_k \tag{5.7}$$

$$\mathbf{P}_{zz,k|k-1} = \int_{R^n} \mathbf{h}(\mathbf{x}_k, \mathbf{u}_k)\mathbf{h}^T(\mathbf{x}_k, \mathbf{u}_k)\mathcal{N}(\mathbf{x}_k; \hat{\mathbf{x}}_{k|k-1}, \mathbf{P}_{k|k-1})d\mathbf{x}_k - \hat{\mathbf{z}}_{k|k-1}\hat{\mathbf{z}}_{k|k-1}^T + \mathbf{R}_k \tag{5.8}$$

and the cross-covariance is

$$\mathbf{P}_{xz,k|k-1} = \int_{R^n} \mathbf{x}_k \mathbf{h}^T(\mathbf{x}_k, \mathbf{u}_k) \mathcal{N}(\mathbf{x}_k; \hat{\mathbf{x}}_{k|k-1}, \mathbf{P}_{k|k-1}) d\mathbf{x}_k - \hat{\mathbf{x}}_{k|k-1} \hat{\mathbf{z}}_{k|k-1}^T. \tag{5.9}$$

Once the new measurement $\mathbf{z}_k$ is received, the CKF computes the posterior density $p(\mathbf{x}_k|D_k)$ and that can be obtained as

$$p(\mathbf{x}_k|D_k) = \mathcal{N}(\mathbf{x}_k; \hat{\mathbf{x}}_{k|k}, \mathbf{P}_{k|k}), \tag{5.10}$$

where

$$\hat{\mathbf{x}}_{k|k} = \hat{\mathbf{x}}_{k|k-1} + \mathbf{K}_k(\mathbf{z}_k - \hat{\mathbf{z}}_{k|k-1}) \tag{5.11}$$

$$\mathbf{P}_{k|k} = \mathbf{P}_{k|k-1} - \mathbf{K}_k \mathbf{P}_{zz,k|k-1} \mathbf{K}_k^T \tag{5.12}$$

with the Kalman gain given as

$$\mathbf{K}_k = \mathbf{P}_{xz,k|k-1} \mathbf{P}_{zz,k|k-1}^{-1} \tag{5.13}$$

It can be seen that in the above prediction and measurement update equations, the Bayesian filter solution reduces to computing the multi-dimensional integrals, whose integrands are all of the form $nonlinear function \times Gaussian$. The heart of the CKF is to solve (or more accurately, to approximate) the multi-dimensional integrals using cubature rules.

5.2.3 Cubature Rules

The cubature rule to approximate an n-dimensional, weighted Gaussian integral is that

$$\int_{R^n} \mathbf{f}(\mathbf{x}) \mathcal{N}(\mathbf{x}; \mu, \mathbf{P}) d\mathbf{x} \approx \frac{1}{2n} \sum_{i=1}^{2n} \mathbf{f}(\mu + \mathbf{P}^{\frac{1}{2}} \xi_i) \tag{5.14}$$

where $\mathbf{P}^{\frac{1}{2}}$ is a square-root factor of the covariance $\mathbf{P}$ satisfying the relation $\mathbf{P} = \mathbf{P}^{\frac{1}{2}}(\mathbf{P}^{\frac{1}{2}})^T$; the set of $2n$ cubature points are given by $\{\xi_i\}$ where ξ_i is the $i\text{-}th$ element of the following set:

$$\sqrt{n} \left[\begin{pmatrix} 1 \\ 0 \\ \vdots \\ 0 \end{pmatrix}, \ldots, \begin{pmatrix} 0 \\ \vdots \\ 0 \\ 1 \end{pmatrix}, \begin{pmatrix} -1 \\ 0 \\ \vdots \\ 0 \end{pmatrix}, \ldots, \begin{pmatrix} 0 \\ \vdots \\ 0 \\ -1 \end{pmatrix} \right] \tag{5.15}$$

These cubature rules are required to numerically evaluate the multi-integrands in the prediction and update stages of the CKF.

The cubature points required for the prediction stage are

$$\boldsymbol{\chi}_{i,k|k-1} = \mathbf{P}_{k-1|k-1}^{\frac{1}{2}}\xi_i + \hat{\mathbf{x}}_{k-1|k-1} \tag{5.16}$$

where $i = 1, 2, \ldots, 2n$ and n is the size of the state vector.

The propagated cubature points through the process model are

$$\boldsymbol{\chi}_{i,k|k-1}^{*} = \mathbf{f}(\boldsymbol{\chi}_{i,k-1|k-1}, \mathbf{u}_{k-1}). \tag{5.17}$$

The predicted mean and error covariance matrix from (5.4), (5.5) and (5.14) are

$$\hat{\mathbf{x}}_{k|k-1} = \frac{1}{2n}\sum_{i=1}^{2n}\boldsymbol{\chi}_{i,k|k-1}^{*} \tag{5.18}$$

$$\mathbf{P}_{k|k-1} = \frac{1}{2n}\sum_{i=1}^{2n}\boldsymbol{\chi}_{i,k|k-1}^{*}\boldsymbol{\chi}_{i,k|k-1}^{*T} - \hat{\mathbf{x}}_{k|k-1}\hat{\mathbf{x}}_{k|k-1}^{T} + \mathbf{Q}_{k-1}. \tag{5.19}$$

By using (5.7)–(5.9) and (5.14), the predicted measurement and its associated covariances are

$$\hat{\mathbf{z}}_{k|k-1} = \frac{1}{2n}\sum_{i=1}^{2n}\mathbf{z}_{i,k|k-1} \tag{5.20}$$

$$\mathbf{P}_{zz,k|k-1} = \frac{1}{2n}\sum_{i=1}^{2n}\mathbf{z}_{i,k|k-1}\mathbf{z}_{i,k|k-1}^{T} - \hat{\mathbf{z}}_{k|k-1}\hat{\mathbf{z}}_{k|k-1}^{T} + \mathbf{R}_k \tag{5.21}$$

$$\mathbf{P}_{xz,k|k-1} = \frac{1}{2n}\sum_{i=1}^{2n}\boldsymbol{\chi}_{i,k|k-1}\mathbf{z}_{i,k|k-1}^{T} - \hat{\mathbf{x}}_{k|k-1}\hat{\mathbf{z}}_{k|k-1}^{T} \tag{5.22}$$

where

$$\mathbf{z}_{i,k|k-1} = \mathbf{h}(\boldsymbol{\chi}_{i,k|k-1}, \mathbf{u}_k) \tag{5.23}$$

$$\boldsymbol{\chi}_{i,k|k-1} = \mathbf{P}_{k|k-1}^{\frac{1}{2}}\xi_i + \hat{\mathbf{x}}_{k|k-1}. \tag{5.24}$$

The updated state and covariance can be obtained by using Eqs. (5.11) and (5.13). Based on the above, the CKF can be summarised in Algorithm 5.1.

5.3 Cubature Transformation

Consider the polar to cartesian nonlinear transformation. The transformation can be described by

$$\begin{bmatrix} x \\ y \end{bmatrix} = \mathbf{h}(\mathbf{x}) = \begin{bmatrix} r\cos\theta \\ r\sin\theta \end{bmatrix} \tag{5.25}$$

where $\mathbf{x}$, consisting of polar coordinates is $\mathbf{x} = [r \quad \theta]^T$, and $[x \quad y]^T$ are the cartesian coordinates, the outcome of the transformation. It is assumed that r and θ are two

---

Algorithm 5.1 Cubature Kalman filter

Initialise the state vector, $\hat{\mathbf{x}}_{0|0}$, and the covariance matrix, $\mathbf{P}_{0|0}$ (set $k = 1$).

Prediction

1: Factorise the covariance matrix, $\mathbf{P}_{k-1|k-1}$

$$\mathbf{P}_{k-1|k-1} = \mathbf{P}^{\frac{1}{2}}_{k-1|k-1}\mathbf{P}^{\frac{T}{2}}_{k-1,k-1}$$

where $\mathbf{P}^{\frac{1}{2}}_{k-1|k-1}$ is the square-root factor of $\mathbf{P}_{k-1|k-1}$.

2: Calculate the cubature points

$$\chi_{i,k-1|k-1} = \mathbf{P}^{\frac{1}{2}}_{k-1|k-1}\xi_i + \hat{\mathbf{x}}_{k-1|k-1}, \quad i = 1, \ldots, 2n$$

where ξ_i is the $i-th$ element of the following set:

$$\sqrt{n}\left[\begin{pmatrix} 1 \\ 0 \\ \vdots \\ 0 \end{pmatrix}, \ldots, \begin{pmatrix} 0 \\ \vdots \\ 0 \\ 1 \end{pmatrix}, \begin{pmatrix} -1 \\ 0 \\ \vdots \\ 0 \end{pmatrix}, \ldots, \begin{pmatrix} 0 \\ \vdots \\ 0 \\ -1 \end{pmatrix}\right].$$

3: Propagate the cubature points through the nonlinear process model

$$\chi^*_{i,k-1|k-1} = \mathbf{f}(\chi_{i,k-1|k-1}, \mathbf{u}_{k-1}), \quad i = 1, \ldots, 2n.$$

4: Predicted state and covariance can be obtained as

$$\hat{\mathbf{x}}_{k|k-1} = \frac{1}{2n}\sum_{i=1}^{2n}\chi^*_{i,k-1|k-1}$$

$$\mathbf{P}_{k|k-1} = \frac{1}{2n}\sum_{i=1}^{2n}\chi^*_{i,k-1|k-1}\chi^{*T}_{i,k-1|k-1} - \hat{\mathbf{x}}_{k|k-1}\hat{\mathbf{x}}^T_{k|k-1} + \mathbf{Q}_{k-1}.$$

Measurement update

1: Factorise the predicted covariance, $\mathbf{P}_{k|k-1}$

$$\mathbf{P}_{k|k-1} = \mathbf{P}^{\frac{1}{2}}_{k|k-1}\mathbf{P}^{\frac{T}{2}}_{k|k-1}.$$

2: Evaluate the cubature points (these cubature points are calculated using predicted mean and covariance, $\hat{\mathbf{x}}_{k|k-1}$ and $\mathbf{P}_{k|k-1}$)

$$\chi_{i,k|k-1} = \mathbf{P}^{\frac{1}{2}}_{k|k-1}\xi_i + \hat{\mathbf{x}}_{k|k-1}.$$

3: Propagate the cubature points through the nonlinear measurement model

$$\mathbf{z}_{i,k|k-1} = \mathbf{h}(\chi_{i,k|k-1}, \mathbf{u}_k).$$

4: Predicted measurement, covariance and cross-covariance can be calculated as

$$\hat{\mathbf{z}}_{k|k-1} = \frac{1}{2n}\sum_{i=1}^{2n} \mathbf{z}_{i,k|k-1}$$

$$\mathbf{P}_{zz,k|k-1} = \frac{1}{2n}\sum_{i=1}^{2n} \mathbf{z}_{i,k|k-1}\mathbf{z}^T_{i,k|k-1} - \hat{\mathbf{z}}_{k|k-1}\hat{\mathbf{z}}^T_{k|k-1} + \mathbf{R}_k$$

$$\mathbf{P}_{xz,k|k-1} = \frac{1}{2n}\sum_{i=1}^{2n} \chi_{i,k|k-1}\mathbf{z}^T_{i,k|k-1} - \hat{\mathbf{x}}_{k|k-1}\hat{\mathbf{z}}^T_{k|k-1}$$

where the Kalman Gain is

$$\mathbf{K}_k = \mathbf{P}_{xz,k|k-1}\mathbf{P}^{-1}_{zz,k|k-1}.$$

5: Update mean and error covariance can be obtained as

$$\hat{\mathbf{x}}_{k|k} = \hat{\mathbf{x}}_{k|k-1} + \mathbf{K}_k(\mathbf{z}_k - \hat{\mathbf{z}}^T_{k|k-1})$$

$$\mathbf{P}_{k|k} = \mathbf{P}_{k|k-1} - \mathbf{K}_k\mathbf{P}_{zz,k|k-1}\mathbf{K}^T_k.$$

---

independent variables with means $\bar{r}$ and $\bar{\theta}$, and the corresponding standard deviations are σ_r and σ_θ, respectively.

Consider a nonlinear function given by $\mathbf{s} = \mathbf{h}(\mathbf{x})$ where $\mathbf{x} \in \mathbb{R}^n$. The mean and covariance of the random variable $\mathbf{x}$ are $\bar{\mathbf{x}}$ and $\mathbf{P}_\mathbf{x}$, respectively. The mean and covariance of $\mathbf{s}$, $\bar{\mathbf{s}}$ and $\mathbf{P}_s$, using cubature transformation can be calculated using Algorithm 5.2.

Algorithm 5.2 Cubature Transform

1: Compute the $2n$ cubature points

$$\chi_i = \bar{\mathbf{x}} + \mathbf{P}_x^{\frac{1}{2}} \xi_i, \quad i = 1, \ldots, 2n \tag{5.26}$$

where $\mathbf{P}^{\frac{1}{2}}$ is the square-root factor or $\mathbf{P}$ and ξ_i is the ith column of

$$\sqrt{n}\left[\begin{pmatrix}1\\0\\\vdots\\\vdots\\0\end{pmatrix}, \ldots, \begin{pmatrix}0\\\vdots\\\vdots\\0\\1\end{pmatrix}, \begin{pmatrix}-1\\0\\\vdots\\\vdots\\0\end{pmatrix}, \ldots, \begin{pmatrix}0\\\vdots\\\vdots\\0\\-1\end{pmatrix}\right]. \tag{5.27}$$

The corresponding weights are

$$W_i = \frac{1}{2n}, \quad i = 1, \ldots, 2n \tag{5.28}$$

2: Propagate the cubature points through the nonlinear function

$$\mathfrak{s}_i = \mathbf{h}(\chi_i), \quad i = 1, \ldots, 2n. \tag{5.29}$$

3: The mean and covariance for $\mathbf{s}$ can be calculated as

$$\bar{\mathbf{s}} \approx \sum_{i=0}^{2n} W_i \mathfrak{s}_i, \quad i = 0, \ldots, 2n \tag{5.30}$$

$$\mathbf{P}_s \approx \sum_{i=0}^{2n} W_i (\mathfrak{s}_i \mathfrak{s}_i^T - \bar{\mathbf{s}}\bar{\mathbf{s}}^T), \quad i = 0, \ldots, 2n. \tag{5.31}$$

5.3.1 Polar to Cartesian Coordinate Transformation—Cubature Transformation

In this subsection, the polar to cartesian coordinate transformation considered in Sect. 4.6 will be analysed again by using the cubature transformation. The cubature transformation given in Algorithm 5.2 will be used to obtain the mean and covariance of Eq. (5.25). The parameters in this subsection are the same as given in Sect. 4.6.

The cubature points using Eq. (5.26) can be calculated as

$$\begin{aligned}
\chi_1 = \bar{\mathbf{x}} + \mathbf{P}_x^{\frac{1}{2}} \xi_1 &= \bar{\mathbf{x}} + \left[\sqrt{n}\mathbf{P}_\mathbf{x}^{\frac{1}{2}}\right]_1 &= \begin{bmatrix} 1 + \sigma_r \sqrt{n} \\ \frac{\pi}{2} \end{bmatrix} \\
\chi_2 = \bar{\mathbf{x}} + \mathbf{P}_x^{\frac{1}{2}} \xi_2 &= \bar{\mathbf{x}} + \left[\sqrt{n}\mathbf{P}_\mathbf{x}^{\frac{1}{2}}\right]_2 &= \begin{bmatrix} 1 \\ \frac{\pi}{2} + \sigma_\theta \sqrt{n} \end{bmatrix} \\
\chi_3 = \bar{\mathbf{x}} + \mathbf{P}_x^{\frac{1}{2}} \xi_3 &= \bar{\mathbf{x}} - \left[\sqrt{n}\mathbf{P}_\mathbf{x}^{\frac{1}{2}}\right]_3 &= \begin{bmatrix} 1 - \sigma_r \sqrt{n} \\ \frac{\pi}{2} \end{bmatrix}
\end{aligned}$$

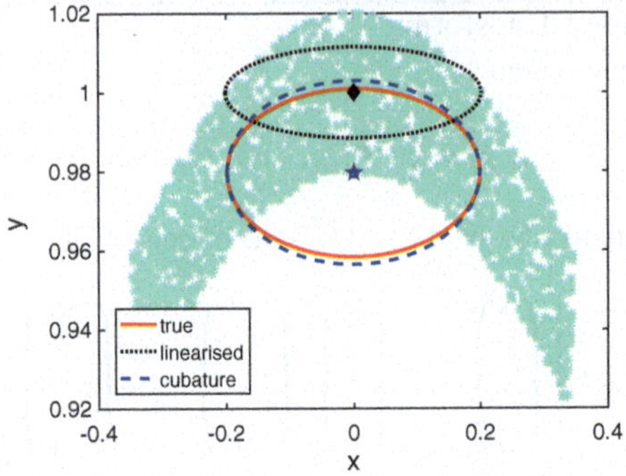

Fig. 5.1 2000 random measurements are generated with range and bearings, which are uniformly distributed between ± 0.02 and $\pm 20°$. These random points are then processed through the nonlinear polar to cartesian coordinate transformation and are shown as $*$. The true-, linearised-, and cubature transformation means are represented by $\bullet$, $\blacklozenge$, and $\star$, respectively. True, linearised, and cubature transformation uncertainty ellipses are represented by solid, dotted, dashed-dotted and dashed lines, respectively

$$\chi_4 = \bar{\mathbf{x}} + \mathbf{P}_x^{\frac{1}{2}} \xi_4 \quad = \quad \bar{\mathbf{x}} - \left[\sqrt{n}\mathbf{P}_{\mathbf{x}}^{\frac{1}{2}}\right]_4 \quad = \quad \begin{bmatrix} 1 \\ \frac{\pi}{2} - \sigma_\theta \sqrt{n} \end{bmatrix} \tag{5.32}$$

The transformed cubature points using Eq. (5.29) are

$$\begin{aligned}
\mathfrak{s}_1 &= \mathbf{h}(\chi_1) = \mathbf{h}\left(\begin{bmatrix} 1 + \sigma_r \sqrt{n} \\ \frac{\pi}{2} \end{bmatrix}\right) = \begin{bmatrix} 0 \\ 1 + \sigma_r \sqrt{(n)} \end{bmatrix} \\
\mathfrak{s}_2 &= \mathbf{h}(\chi_2) = \mathbf{h}\left(\begin{bmatrix} 1 \\ \frac{\pi}{2} + \sigma_\theta \sqrt{n} \end{bmatrix}\right) = \begin{bmatrix} \cos\left(\frac{\pi}{2} + \sigma_\theta \sqrt{(n)}\right) \\ \sin\left(\frac{\pi}{2} + \sigma_\theta \sqrt{(n)}\right) \end{bmatrix} \\
\mathfrak{s}_3 &= \mathbf{h}(\chi_3) = \mathbf{h}\left(\begin{bmatrix} 1 \\ \frac{\pi}{2} + \sigma_\theta \sqrt{n} \end{bmatrix}\right) = \begin{bmatrix} 0 \\ 1 - \sigma_r \sqrt{(n+)} \end{bmatrix} \\
\mathfrak{s}_4 &= \mathbf{h}(\chi_4) = \mathbf{h}\left(\begin{bmatrix} 1 \\ \frac{\pi}{2} - \sigma_\theta \sqrt{n} \end{bmatrix}\right) = \begin{bmatrix} \cos\left(\frac{\pi}{2} - \sigma_\theta \sqrt{(n)}\right) \\ \sin\left(\frac{\pi}{2} - \sigma_\theta \sqrt{(n)}\right) \end{bmatrix}.
\end{aligned} \tag{5.33}$$

Once the cubature points and its corresponding weights, and the transformed cubature points are obtained; the mean and covariance of $\mathbf{s}$ can be calculated as

$$\bar{\mathfrak{s}} \approx \sum_{i=1}^{4} W_i \mathfrak{s}_i \tag{5.34}$$

$$\mathbf{P}_s \approx \sum_{i=1}^{4} W_i (\mathfrak{s}_i \mathfrak{s}_i^T - \bar{\mathfrak{s}}\bar{\mathfrak{s}}^T). \tag{5.35}$$

Simulations in Sect. 4.6 have been repeated along with cubature transformation and the corresponding results are shown in Fig. 5.1. The true mean is at (0, 0.9798), the linearised mean is at (0, 1), and the cubature mean is at (0, 0.9798). It is very hard to see the means of different transformations in Fig. 5.1 as they are overlapped. However, by zooming in the area around the means, they can be clearly distinguished. The uncertainty ellipse using cubature transformation is very close to the true ellipse.

5.4 Study on a Brushless DC Motor

In this section a case study of a brushless DC motor (BLDC motor)is to be considered. It will show how a cubature Kalman filter is used to estimate the unmeasurable states of the motor and the estimated states are to be fed back to the various components of the BLDC drive to control the speed of the motor. This section first presents a detailed nonlinear model of the BLDC motor (Lee and Ehsani 2003). The state vector of BLDC motor consists of currents, the speed and rotor position. The basic block diagram consists of BLDC motor, EMF block, inverter and hysteresis block, and a PID controller is shown in Fig. 5.2 (Potnuru et al. 2016). The inputs to the BLDC motor block are voltages, back EMFs and load torque and the output consists of the full state vector. The reference speed and the actual speed are compared and the error between them is controlled by the PID controller, the output of the PID controller is required to obtain the reference current, I_{ref}. Part of the inputs to the BLDC motor are obtained from inverter subsystem and the other part, the back EMF, can be calculated from the EMF block. Each of these blocks is detailed in the next few subsections.

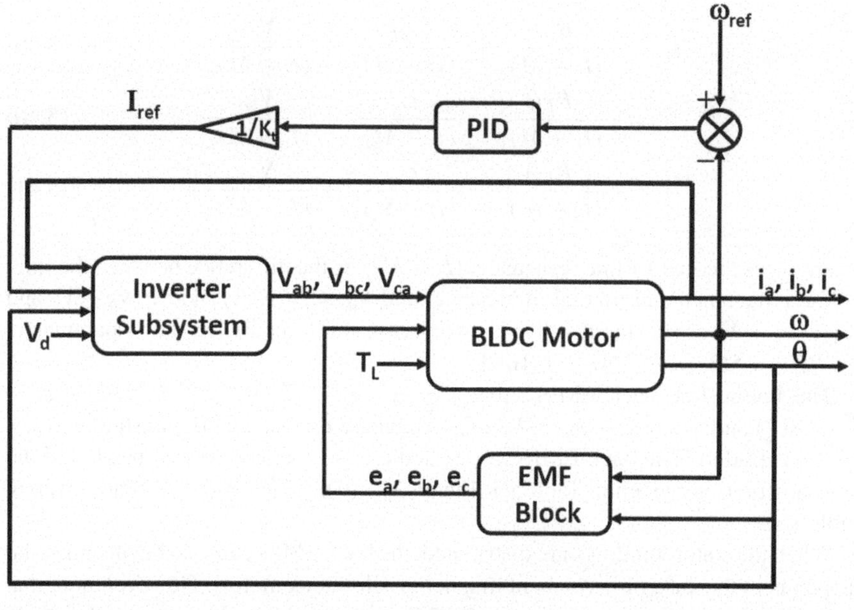

Fig. 5.2 Block diagram of full state feedback BLDC motor

Table 5.1 Back EMFs as functions of θ and ω

	e_a	e_b	e_c
$0 \le \theta < \frac{\pi}{6}$	$\frac{6K_e\omega}{\pi}\theta$	$-K_e\omega$	$K_e\omega$
$\frac{\pi}{6} \le \theta < \frac{\pi}{2}$	$K_e\omega$	$-K_e\omega$	$K_e\omega\left[-\frac{6}{\pi}\theta + 2\right]$
$\frac{\pi}{2} \le \theta < \frac{5\pi}{6}$	$K_e\omega$	$K_e\omega\left[\frac{6}{\pi}\theta - 4\right]$	$-K_e\omega$
$\frac{5\pi}{6} \le \theta < \frac{7\pi}{6}$	$K_e\omega\left[-\frac{6}{\pi}\theta + 6\right]$	$K_e\omega$	$-K_e\omega$
$\frac{7\pi}{6} \le \theta < \frac{9\pi}{6}$	$-K_e\omega$	$K_e\omega$	$K_e\omega\left[\frac{6}{\pi}\theta - 8\right]$
$\frac{9\pi}{6} \le \theta < \frac{11\pi}{6}$	$-K_e\omega$	$K_e\omega\left[-\frac{6}{\pi}\theta + 10\right]$	$K_e\omega$
$\frac{11\pi}{6} \le \theta \le 2\pi$	$K_e\omega\left[\frac{6}{\pi}\theta - 12\right]$	$K_e\omega$	$K_e\omega$

5.4.1 BLDC Motor Dynamics

The state vector of BLDC motor, $[i_a \;\; i_b \;\; i_c \;\; \omega \;\; \theta]^T$, consists of phase currents, the speed and the rotor position. The phase currents can be obtained as

$$i_a = i_1 - i_3 \tag{5.36}$$
$$i_b = i_2 - i_1 \tag{5.37}$$
$$i_c = i_3 - i_2, \tag{5.38}$$

where the line-to-line currents, i_1, i_2 and i_3, can be obtained by solving the below dynamical equations (Lee and Ehsani 2003)

$$\dot{i}_1 = -\frac{R_L i_1}{(L-M)_L} - \frac{e_{ab}}{(L-M)_L} + \frac{\mathbf{V}_{ab}}{(L-M)_L} \tag{5.39}$$
$$\dot{i}_2 = -\frac{R_L i_2}{(L-M)_L} - \frac{e_{bc}}{(L-M)_L} + \frac{\mathbf{V}_{bc}}{(L-M)_L} \tag{5.40}$$
$$\dot{i}_3 = -\frac{R_L i_3}{(L-M)_L} - \frac{e_{ca}}{(L-M)_L} + \frac{\mathbf{V}_{ca}}{(L-M)_L}, \tag{5.41}$$

where R_L is the line-to-line resistance, $(L - M)_L$ is the difference between the line-to-line inductance and mutual inductance; and $V_{ab} = V_a - V_b$, $V_{bc} = V_b - V_c$ and $V_{ca} = V_c - V_a$ are the input voltages. In this case study, the BLDC motor parameters are $R_L = 1.5\Omega$, $(L - M)_L = 6.1$mH.

The line-to-line back EMFs are $e_{ab} = e_a - e_b = K_e\omega f_{ab}(\theta)$, $e_{bc} = e_b - e_c = K_e\omega f_{bc}(\theta)$ and $e_{ca} = e_c - e_a = K_e\omega f_{ab}(\theta)$, where the back EMF constant is $K_e = 0.1074V/$(rad/s). The back EMFs, e_a, e_b and e_c, as functions of the speed; and the rotor positions are given in Table 5.1. Finally, $f_{ab}(\theta)$, $f_{bc}(\theta)$ and $f_{ca}(\theta)$ are given in Table 5.2.

When the stator windings are distributed, the back EMFs can also be of sinusoidal shape (as in the case of permanent magnet synchronous motor). However, note that in this study a tetra square-wave type BLDC motor is used to demonstrate the effectiveness of the algorithms and hence the details in Table 5.1 are given for trapezoidal waveforms. The speed and the rotor position are

$$\dot{\omega} = \frac{1}{J}\left[T_e - T_L\right] \tag{5.42}$$

$$\dot{\theta} = \frac{P}{2}\omega, \tag{5.43}$$

where $P = 6$ is the number of poles and the inertia $J = 8.2614 \times 10^{-5}\text{kgm}^2$, T_L is the load torque and the electro-magnetic torque T_e is

$$T_e = \frac{e_a i_a + e_b i_b + e_c i_c}{\omega}. \tag{5.44}$$

At the first instant, the above equations seem to be linear state equations, but actually they are not. This is mainly due to the back EMFs, which involves several nonlinearities and will be discussed in Sect. 5.4.2. Further it might look like the electro-magnetic torque (T_e) will be infinity if the speed ω is zero. However, after expanding the back EMF terms (e_a, e_b and e_c) as given in Table 5.1, the speed in the numerator and denominator of T_e are cancelled. For example, consider the case when the $0 \le \theta \le \frac{\pi}{6}$, $e_a = \frac{6K_e\omega}{\pi\theta}$, $e_b = -K_e\omega$, $e_c = K_e\omega$. Once these back EMFs are substituted in the T_e expression it yields $T_e = \frac{6K_e}{\pi\theta}i_a - K_e i_b + K_e i_c$, and hence T_e is independent of ω.

Note that in this book, the estimation is conducted with discrete-time models. The continuous BLDC motor model has been discretised using the Euler's method with sampling frequency of 10000 Hz.

5.4.2 Back EMFs and PID Controller

This subsection describes the back EMFs and PID controller design.

Back EMFs

The evaluation of back EMFs plays a key role in the overall BLDC motor operation. They mainly depend on the speed and the rotor position (Lee and Ehsani 2003) and can be calculated using Table 5.1.

PID Controller

The main aim of the PID controller is to bring the actual speed as close as to the reference speed. It is impossible to achieve the desired performance if the PID gains are not properly tuned. In the study presented in this book, the desired closed loop performance in terms of time domain specifications is chosen as the settling time $< 1s$ and peak overshoot $< 20\%$. It is possible to choose various combinations of PID gains to achieve the desired performance. An attempt has been made to design

Table 5.2 The functions f_{ab}, f_{bc} and f_{ca}

	f_{ab}	f_{bc}	f_{ca}
$0 \le \theta < \frac{\pi}{6}$	$\frac{6\theta}{\pi} + 1$	-2	$-\frac{6\theta}{\pi} + 1$
$\frac{\pi}{6} \le \theta < \frac{\pi}{2}$	2	$\frac{6\theta}{\pi} - 5$	$3 - \frac{6\theta}{\pi}$
$\frac{\pi}{2} \le \theta < \frac{5\pi}{6}$	$-\frac{6\theta}{\pi} + 5$	$\frac{6\theta}{\pi} - 3$	-2
$\frac{5\pi}{6} \le \theta < \frac{7\pi}{6}$	$-\frac{6\theta}{\pi} + 5$	2	$\frac{6\theta}{\pi} - 7$
$\frac{7\pi}{6} \le \theta < \frac{9\pi}{6}$	-2	$-\frac{6\theta}{\pi} + 9$	$\frac{6\theta}{\pi} - 7$
$\frac{9\pi}{6} \le \theta < \frac{11\pi}{6}$	$\frac{6\theta}{\pi} - 11$	$-\frac{6\theta}{\pi} + 9$	2
$\frac{11\pi}{6} \le \theta \le 2\pi$	$\frac{6\theta}{\pi} - 11$	-2	$13 - \frac{6\theta}{\pi}$

and implement PID controller. And, it has been found that a PI controller (with zero derivative gain) can adequately achieve the desired performance. The Ziegler–Nichols method has been employed to find the PI controller gains, which were then manually fine tuned to obtain the desired response. The tuned gains are, $K_p = 0.05$ and $K_i = 10$. The output of the PI controller is scaled by K_t to obtain the reference current I_{ref}, which is required for the inverter subsystem to generate the desired input voltages. In this study the torque constant is assumed to be $K_t = 0.21476\,\text{Nm}$.

Inverter and Hysteresis Block

The inverter subsystem is responsible to generate the required voltages for the BLDC motor based on the hysteresis current control. The hysteresis current control provides the switching functions which are required to calculate the phase and line-to-line voltages. The switching functions can be obtained using Table 5.3, where I_{ad}, I_{bd} and I_{cd} are the delayed phase currents (Lee and Ehsani 2003), and T_1 and T_2 depends on the corresponding current sequence.

For example, consider the case of current i_a. When i_a is positive

$$T_1 = \Big[(i_a < 0.9I_{ref}) - (i_a > 1.1I_{ref}) + (i_a > 0.9I_{ref}) \times (i_a < 1.1I_{ref}) \times (i_a > i_{ad}) \\ -(i_a > 0.9I_{ref}) \times (i_a < 1.1I_{ref}) \times (i_a < i_{ad})\Big] \tag{5.45}$$

The functioning of T_1 is based on the i_a, I_{ref} and the previous sample current i_{ad}. Suppose if $(i_a < 0.9I_{ref})$ i.e. the first term of the above equation is satisfied (and

Table 5.3 Switching functions for hysteresis current control

Inputs	Outputs
I_a, I_{ad}	$F_a = (\theta > \pi/6)(\theta < 5\pi/6)T_1 + (\theta > 7\pi/6)(\theta < 11\pi/6)T_2$
I_b, I_{bd}	$F_b = (\theta > 0)(\theta < \pi/2)T_2 + (\theta > 5\pi/6)(\theta < 9\pi/6)T_1$
	$+(\theta > 11\pi/6)(\theta < 2\pi)T_2$
I_c, I_{cd}	$F_c = (\theta > 0)(\theta < \pi/6)T_1 + (\theta > \pi/2)(\theta < 7\pi/6)T_2$ $+(\theta > 9\pi/6)(\theta < 2\pi)T_1$

hence the other three terms are zero), will give $T_1 = 1$. Similarly, if $i_a > 1.1I_{ref}$, the second term will be activated, which will give $T_1 = -1$. If $0.9I_{ref} < i_a < 1.1I_{ref}$ and $i_a > i_{ad}$ then the third term will be activated, which will make $T = 1$. Finally, if $0.9I_{ref} < i_a < 1.1I_{ref}$ and $i_a < i_{ad}$, then fourth term will be activated to yield $T_1 = -1$.

When i_a is negative,

$$T_2 = \Big[-(i_a > -0.9I_{ref}) + (i_a < -1.1I_{ref}) - (i_a < -0.9I_{ref})(i_a > -1.1I_{ref}) \times (i_a < i_{ad}) \\ +(i_a < -0.9I_{ref}) \times (i_a > -1.1I_{ref}) \times (i_a > i_{ad}) \Big]. \tag{5.46}$$

The lower and upper limits of hysteresis current controller are given by $0.9I_{ref}$ and $1.1I_{ref}$, respectively. Similarly, for the other currents i_b and i_c, the corresponding T_1 and T_2 can be calculated.

Based on these switching functions, the phase voltages can be calculated as

$$\begin{aligned} V_a &= \frac{V_d}{2} F_a \\ V_b &= \frac{V_d}{2} F_b \\ V_c &= \frac{V_d}{2} F_c, \end{aligned} \tag{5.47}$$

where, V_d is the DC link voltage and can be obtained either by pre-design analysis or by using a suitable rectifier. The inverter line-to-line voltages can now be obtained as

$$\begin{aligned} V_{ab} &= V_a - V_b \\ V_{bc} &= V_b - V_c \\ V_{ca} &= V_c - V_a. \end{aligned} \tag{5.48}$$

These line-to-line voltages are the inputs to the BLDC motor (Fig. 5.3).

Finally, the state vector and output of BLDC motor model can be expressed in the state-space form as

$$\dot{\mathbf{x}} = \mathbf{F}(\mathbf{x})\mathbf{x} \quad + \quad \mathbf{Gd} \quad + \quad \mathbf{V}T_L$$

$$\begin{bmatrix} \dot{i}_a \\ \dot{i}_b \\ \dot{i}_c \\ \dot{\omega} \\ \dot{\theta} \end{bmatrix} = \begin{bmatrix} -\frac{R_L}{(L-M)_L} & 0 & 0 & -\frac{K_e}{(L-M)_L} f_{ab}(\theta) & 0 \\ 0 & -\frac{R_L}{(L-M)_L} & 0 & -\frac{K_e}{(L-M)_L} f_{bc}(\theta) & 0 \\ 0 & 0 & -\frac{R_L}{(L-M)_L} & -\frac{K_e}{(L-M)_L} f_{ca}(\theta) & 0 \\ \frac{K_e}{J} f_{ab}(\theta) & \frac{K_e}{J} f_{bc}(\theta) & \frac{K_e}{J} f_{ca}(\theta) & -\frac{B}{J} & 0 \\ 0 & 0 & 0 & \frac{P}{2} & 0 \end{bmatrix} \begin{bmatrix} i_a \\ i_b \\ i_c \\ \omega \\ \theta \end{bmatrix}$$

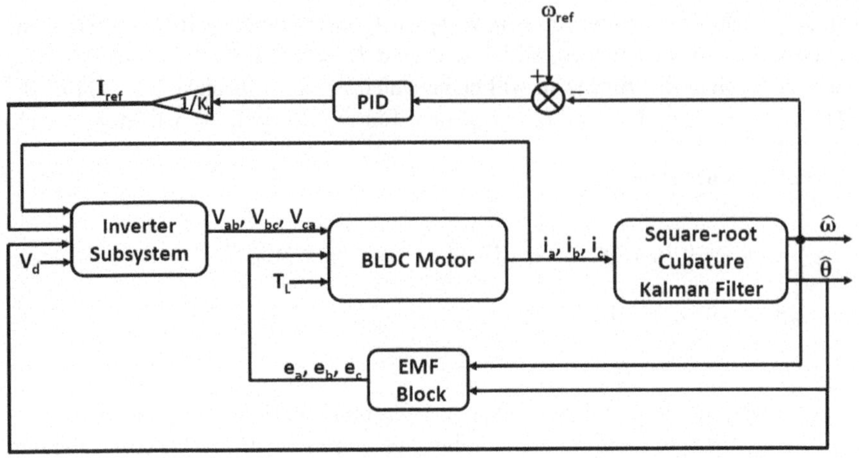

Fig. 5.3 Proposed scheme for a BLDC motor

$$+\begin{bmatrix} \frac{1}{(L-M)_L} & 0 & 0 \\ 0 & \frac{1}{(L-M)_L} & 0 \\ 0 & 0 & \frac{1}{(L-M)_L} \\ 0 & 0 & 0 \\ 0 & 0 & 0 \end{bmatrix} \begin{bmatrix} V_{ab} \\ V_{bc} \\ V_{ca} \end{bmatrix} + \begin{bmatrix} 0 \\ 0 \\ 0 \\ \frac{1}{J} \\ 0 \end{bmatrix} T_L \tag{5.49}$$

The output equation is:

$$\begin{aligned} \mathbf{y} &= \mathbf{H}\mathbf{x} \\ &= \begin{bmatrix} 1\,0\,0\,0\,0 \\ 0\,1\,0\,0\,0 \\ 0\,0\,1\,0\,0 \end{bmatrix} \begin{bmatrix} i_a \\ i_b \\ i_c \\ \omega \\ \theta \end{bmatrix} \end{aligned} \tag{5.50}$$

The process and measurement noises are added to the state and output equations, respectively. The equations in (5.49) and (5.50) are in the continuous-time domain, however the estimation methods implemented are in the discrete-time domain. Hence these equations have been discretised by using Euler's method[1] such that:

$$\mathbf{x}_{n+1} = (\mathbf{I}_5 + \mathbf{F}(\mathbf{x}_n)\Delta t)\mathbf{x}_n + \mathbf{G}\Delta t\mathbf{d} + \mathbf{V}\Delta t T_L \tag{5.51}$$

[1]The continuous state-space model, $\dot{\mathbf{x}} = A\mathbf{x} + B\mathbf{u}$, can be discretised by using Euler's method as

$$\mathbf{x}_{n+1} = (\mathbf{I}_n + A\Delta t)\mathbf{x}_n + \Delta t B\mathbf{u},$$

where Δt is the sampling time and $\mathbf{I}_n$ is the nth order identity matrix where $\mathbf{I}_5$ is the fifth-order identity matrix and Δt is the sampling time. The output equation in the discrete-time domain will be the same as the one given in (5.50), that is

$$\mathbf{y}_n = \mathbf{H}\mathbf{x}_n. \tag{5.52}$$

The following assumptions have been considered for the mathematical model of the BLDC motor and power converters. The magnetic flux in the motor stator winding is not saturated; stator resistances of all three phase windings are equal; self and mutual inductances are constant and they are not varying with load variation; devices in the inverter are idea and iron losses are negligible (Krishnan 2017).

Note that, since the back EMF of a BLDC motor is of trapezoidal shape, the BLDC motor is controlled by exciting only two phases at a time for every 60° of rotor position. Further, one can calculate all the three currents by using only two current sensors as

$$i_a + i_b + i_c = 0.$$

Due to the sensor noise in the current sensors, it is not always possible to accurately calculate all the three currents from two current sensors. Specifically, during the low speed operations the current sensors used in this study have a low signal-to-noise ratio. Hence, three current sensors are used for the efficient control of the BLDC motor (whereas only two phases are excited at a time). However, by using only two high resolution current sensors one can control the BLDC motor efficiently as well.

5.4.3 Initialisation of the State Estimators

In the implementation it is required first to initialise the state vector (five states in all) and the corresponding covariance matrix, and then to choose appropriate process and noise covariance matrices $\mathbf{Q}_n$ and $\mathbf{R}_n$. The initial state vector $\mathbf{x}_0$ and covariance matrix P_0 are randomly selected from $\mathcal{N}([0 \;\; 0 \;\; 0 \;\; 0.01 \;\; 0.01]^T, 10^{-5}\mathbf{I}_5)$, where $\mathcal{N}$ represents the normal-distributed random variables. The process noise covariance $\mathbf{Q}_n$ is a fifth-order diagonal matrix with all the diagonal elements as 10^{-12} and the measurement noise covariance $\mathbf{R}_n$ is a third-order diagonal matrix with all the diagonal elements as 0.01. With the initial values and the plant dynamic model, Algorithm 4.1 and 5.1 can be implemented.

5.4.4 BLDC Motor Experiments

This subsection describes the setup of a BLDC motor system and shows the experimental results using the cubature Kalman filter (CKF) compared with that from the extended Kalman filter (EKF).

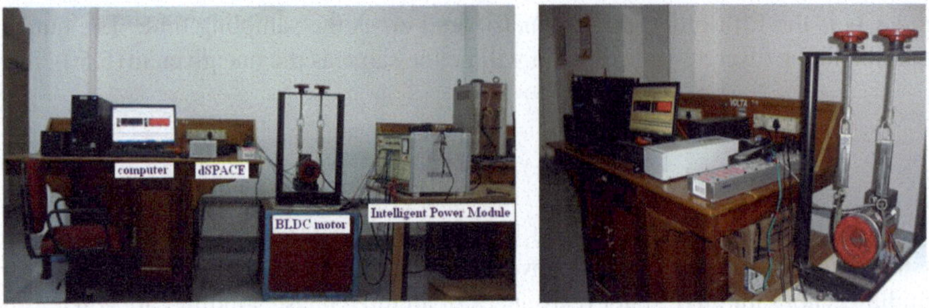

(a) Front view of the experimental setup for BLDC drive (b) Side view of the experimental BLDC drive

Fig. 5.4 Experimental setup for BLDC drive

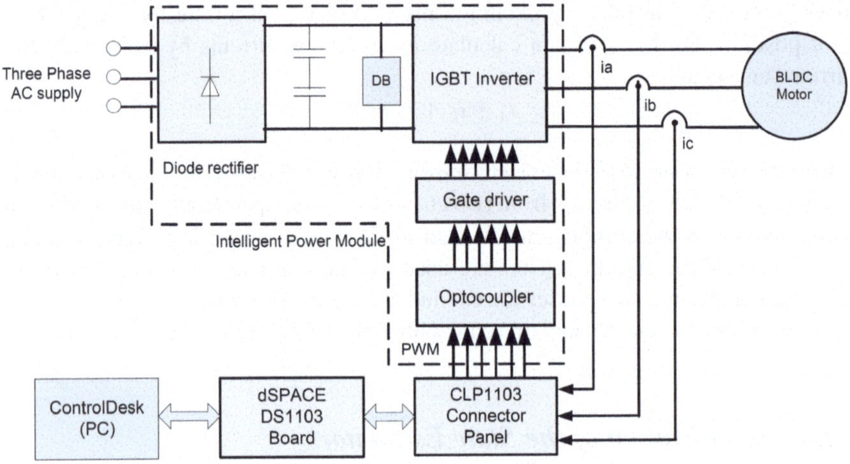

Fig. 5.5 Block diagram for BLDC drive experiment

Experiment Setup

The setup consists of a BLDC motor system with incremental encoder, intelligent power module (MITSUBISHI 1998) (including rectifier, input filter using capacitors, dynamic braking module, voltage source inverter, gate driver circuit and optocoupler), current sensors (LTS 25 NP current transducers), dSPACE DS1103 controller board, dSPACE CLP1103 connector panel (DS1103 2008) and a desk-top computer (dSPACEcontrol desk). These devices are interconnected as shown in Fig. 5.4 and depicted in the block diagram Fig. 5.5.

The measured currents are processed through the dSPACE CLP1103 connector panel which has analog-to-digital and digital-to-analog ports and a provision to give inputs to the dSPACE DS1103 board. The dSPACE DS1103 controller board has a DSP processor which has analog-to-digital and digital-to-analog converters and other components required for the data processing. Several filters are also included for signal conditioning purposes. The block diagram given in Fig. 5.6 is simulated

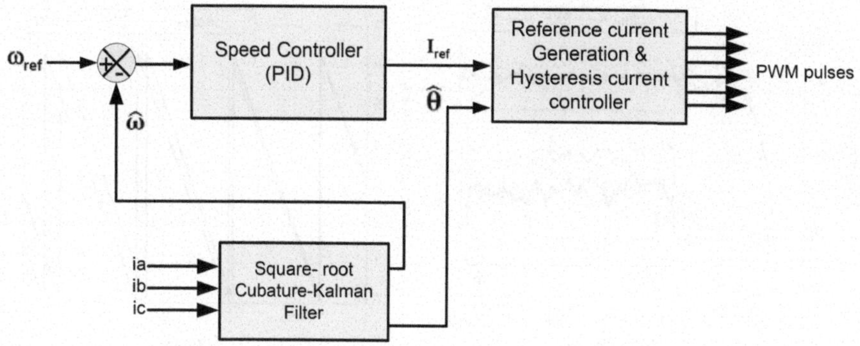

Fig. 5.6 Speed controller scheme using CKF

using the various dSPACE compatible RTI and Simulink blocksets in the computer. This Simulink diagram is then converted to a set of digital signal processing enabled embedded C-code, which is then downloaded to the DSP processor TMS320F240 (in the DS1103 controller board) for further processing to generate the pulse width modulation (PWM) pulses for the inverter. Based on these PWM pulses the inverter generates the required voltage for the BLDC motor.

The block diagram in Fig. 5.5 constitutes a part of the BLDC drive given in Fig. 5.6. The main purpose is to generate the PWM pulses for the inverter. The reference speed and the speed obtained from the square-root CKF block are compared and based on the estimation error the PID block generates the reference current I_{ref}. The I_{ref} along with the estimated rotor position (from the CKF) and measured currents generates the PWM pulses for the inverter (Kim et al. 2007).

Experiment Results

Three sets of experiments have been implemented: one with the full state feedback and the remaining two with estimated states obtained from CKF and EKF, respectively. These experiments are performed for the low reference speed of 30 rpm, high reference speed of 2000 rpm and with the parametric variations. For the full state feedback experiment, all the states are measured using LTS 25 NP current transducers, and the speed and rotor position are measured by an incremental encoder and hall sensors. However, only current sensors are utilised for the BLDC drive using the EKF and CKF approaches.

1. **Low-speed results**
   In this experiment, the reference speed of the BLDC motor is chosen at 30 rpm. The actual speed, and the estimated speed using CKF and EKF are shown in Fig. 5.7a and the corresponding rotor positions are shown in Fig. 5.7b. The measured speed and the estimated speed using CKF are close to each other, whereas the estimated speed using EKF has significant discrepancies with the measured speed. The errors between the measured and estimated speeds are given in Fig. 5.7c. The

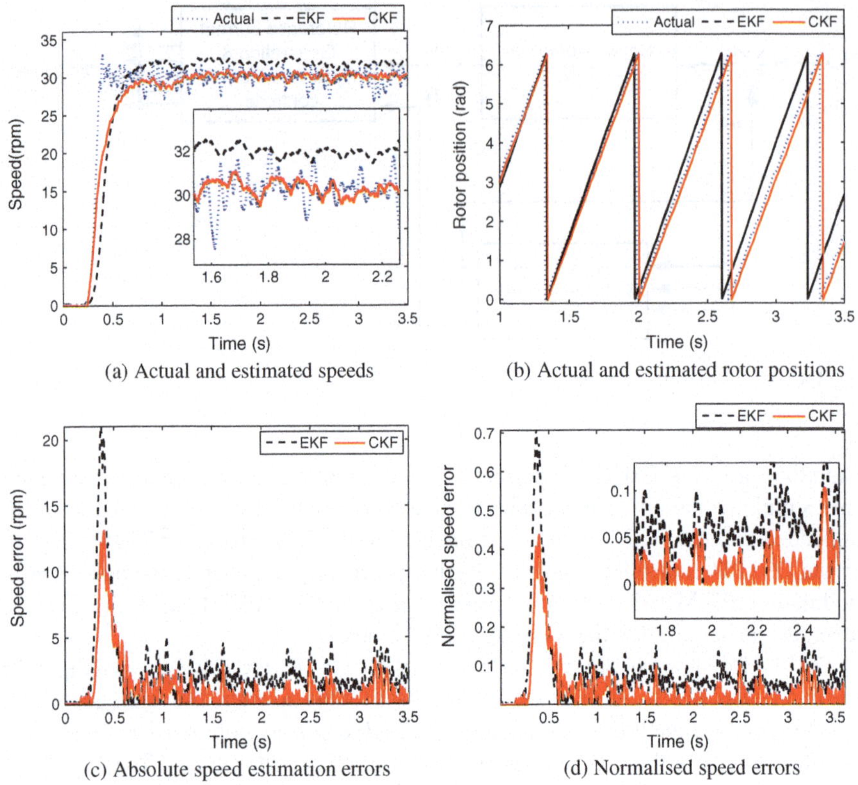

Fig. 5.7 Experimental results for low reference speed (30 rpm)

maximum estimation error for the EKF is around 20 rpm and for the CKF is around 12 rpm. The average absolute speed estimation error is 2.3982 rpm and 1.2525 rpm for EKF and CKF, respectively. The normalised speed error plot is shown in Fig. 5.7d. The normalised speed is a ratio of the speed error and the reference speed.

2. **High-speed results**
   In this experiment, the reference speed of the BLDC motor is chosen as a ramp signal (slope of 400 rpm/s) with the final value of 2000 rpm. The actual speed, and the estimated speed using EKF and CKF are shown in Fig. 5.8a and the corresponding rotor positions are shown in Fig. 5.8b. Similar to the low-speed simulations, the measured speed and the estimated speed using CKF are close to each other, whereas the estimated speed using EKF has significant errors. The errors between the measured and estimated speeds are given in Fig. 5.8c. The maximum estimation error for the EKF is around 115 rpm and for the CKF is around 40 rpm. The average absolute speed estimation error is 25.8198 rpm and 13.1036 rpm for EKF and CKF, respectively. The normalised speed error plot is shown in Fig. 5.8d.

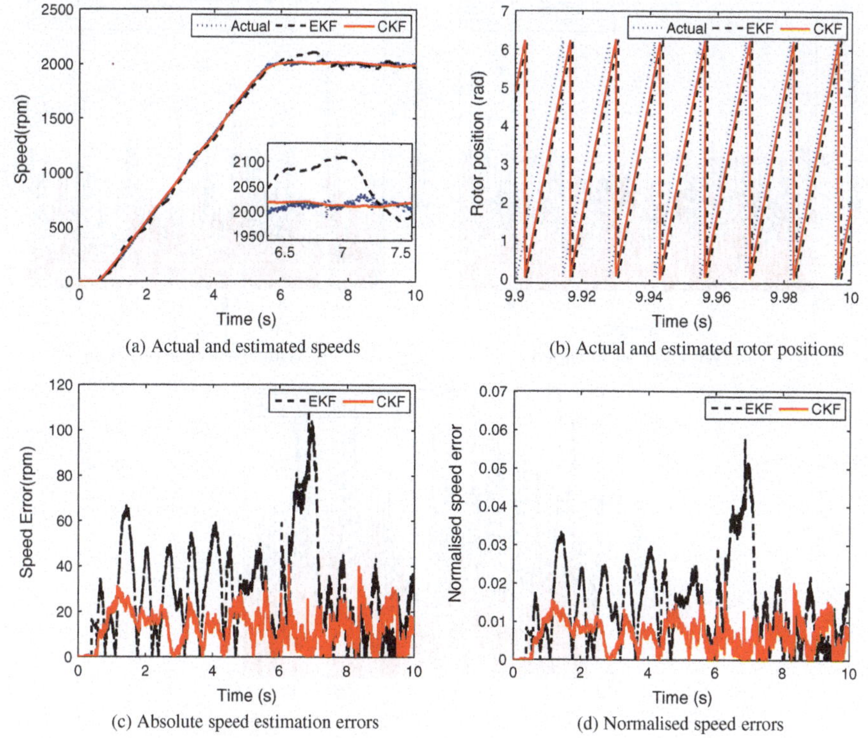

(a) Actual and estimated speeds

(b) Actual and estimated rotor positions

(c) Absolute speed estimation errors

(d) Normalised speed errors

Fig. 5.8 Experimental results for high reference speed (2000 rpm)

3. **Robustness analysis**

   In this part, the performance of the estimators in the presence of various uncertainties is considered. These experiments are performed at the low reference speed of 30 rpm and the steady state results are analysed.

   Load torque variation

   In this experiment, the load torques of 2 N-m and 4 N-m are considered. These mechanical loads are applied from the starting of the experiment. The nominal results (without load torque variation) are given in Fig. 5.9a, the absolute average speed error is 1.7327 rpm and 0.7967 rpm for EKF and CKF, respectively. The speed errors with the load torque of 2 N-m are given in Fig. 5.9b, where the absolute average errors are 1.1968 rpm and 0.8336 rpm for EKF and CKF, respectively. The speed errors with the load torque of 4 N-m are given in Fig. 5.9c, where the average errors for EKF and CKF are 2.4086 rpm and 1.1743 rpm, respectively.

   Stator resistance variation

   In this experiment, the stator resistances for the estimators are varied $+/-30\%$ from the nominal value. The nominal results (without the stator resistance variation) are the same as seen in Fig. 5.9a. The speed errors with $+30\%$ variation

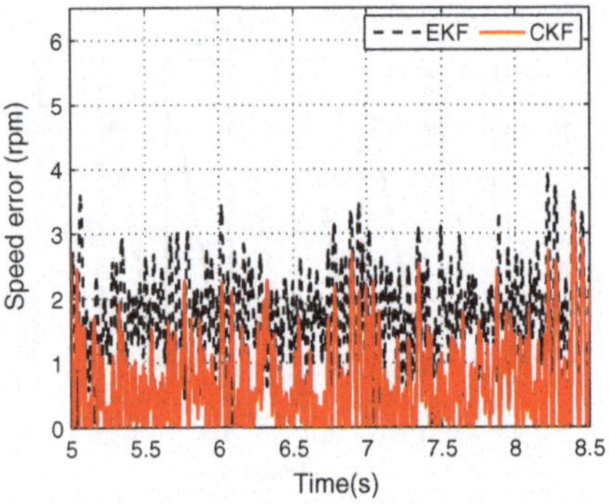

(a) With no load condition

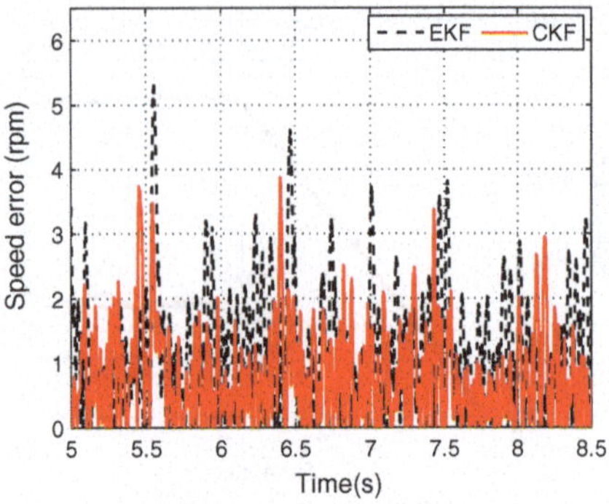

(b) With load torque of 2 N-m

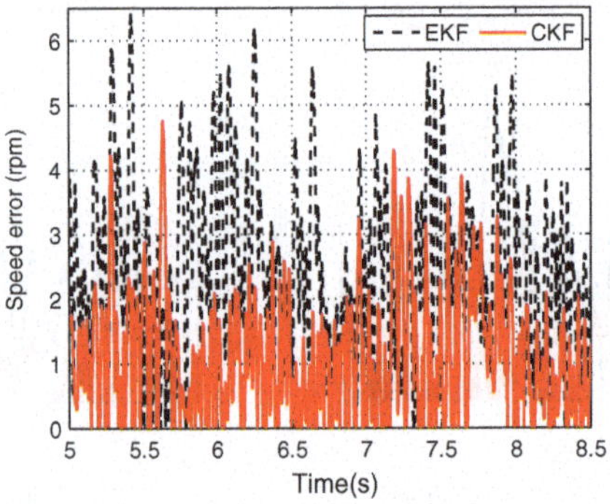

(c) With load torque of 4 N-m

Fig. 5.9 Experimental results with load variations

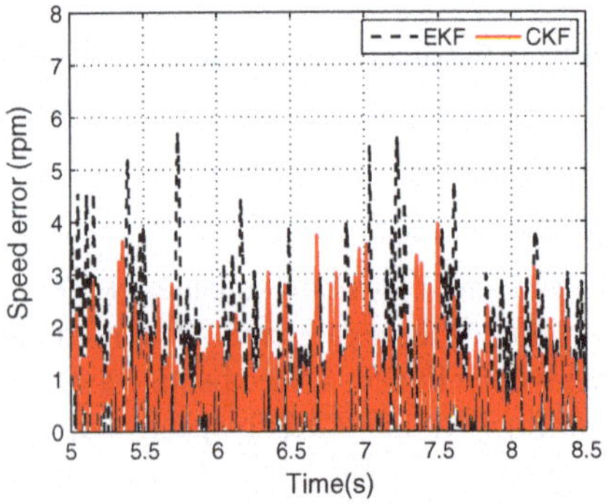

(a) With +30% variation in stator resistance

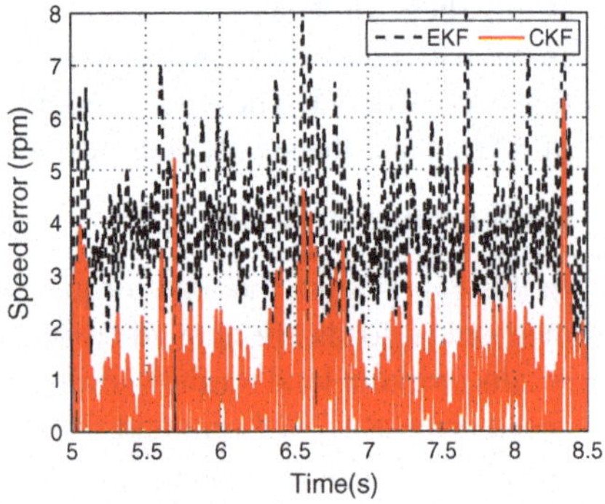

(b) With −30% variation in stator resistance

Fig. 5.10 Experimental results with stator resistance variations

of the stator resistance from its nominal value are given in Fig. 5.10a, where the absolute average errors are 1.9144 rpm and 0.9866 rpm for EKF and CKF, respectively. The speed errors with −30% variation of stator resistance from its nominal

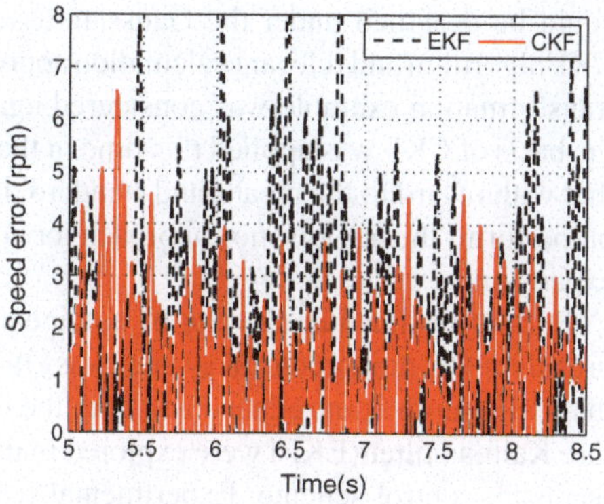

Fig. 5.11 Estimation errors with de-tuned stator inductance

value are given in Fig. 5.10b, where the absolute average errors are 3.1422 rpm and 0.8644 rpm for EKF and CKF, respectively. It can be seen that the speed error is bigger for the negative variation of stator resistance (−30%) as compared to the +30% variation in the stator resistance.

De-tuning of inductance

In this experiment, the stator inductance for the estimators is de-tuned by 10% from its nominal value. The nominal results (without de-tuning) are the same as given in Fig. 5.9a. The results with de-tuned inductance are given in Fig. 5.11, where the absolute average speed errors are 3.5140 rpm and 1.8106 rpm for the cases of EKF and CKF, respectively.

The experimental results shown above demonstrate that it is possible to estimate the speed and rotor positions with either EKF or CKF. The experiments with different parametric variations show the robustness property of the EKF and CKF. However in all the cases the performance of the CKF is better than that of the EKF approach. That is due to the fact that EKF is based on the first-order approximation of the nonlinear function, whereas this approximation is not required for the CKF. Further, the calculation of the Jacobians for the EKF are quite complex and in some cases there is a possibility that these Jacobians can become singular leading to numerical instability. It can also be observed that among the considered uncertainties, the estimator performance is sensitive to the de-tuned inductance.

5.5 Summary

In this chapter the cubature Kalman filter (CKF) has been introduced. CKF does not use the Taylor series expansion of nonlinear functions involved in the system equations, which is obviously a great advantage in state estimation for nonlinear systems. CKF is also the closest known approximation to the Bayesian filter that could be designed under the Gaussian assumption. Following the introduction of CKF algorithm and relevant calculation approaches, the polar to cartesian coordinates transformation example was considered again. The cubature transformation that is the basis of CKF was applied this time in the coordinates transformation. It is shown that with regard to the means and standard deviation ellipses, the cubature transform outperforms the linearisation approach for nonlinear functions using the Taylor series expansion.

Further to illustrate the use of nonlinear state estimation approaches, a case of brushless DC motor control system was studied in this chapter. A laboratory experimental setup was introduced and both the extended Kalman filter (EKF) and cubature Kalman filter (EKF) were explored in the state estimation combined with a state feedback control scheme. Experimental results showed some advantages of using the CKF approach over the EKF approach in such applications.

References

Arasaratnam I, Haykin S (2009) Cubature kalman filters. IEEE Trans Autom Control 54(6):1254–1269

DS1103 PPC Controller Board catalogue, Paderborn, Germany, 2008, howpublished = https://www.dspace.com/en/inc/home/products/hw/singbord/ppcconbo.cfm. Accessed 10 Feb 2018

Kim T, Lee H, Ehsani M (2007) Position sensorless brushless dc motor/generator drives: review and future trends. IET Electr Power Appl 1(4):557–564

Krishnan R (2017) Permanent magnet synchronous and brushless DC motor drives. CRC press, Boca Raton

Lee B-K, Ehsani M (2003) Advanced simulation model for brushless dc motor drives. Electr Power Compon Syst 31(9):841–868

MITSUBISHI Intelligent power modules catologue, 1998, howpublished = http://www.mitsubishielectric.com/semiconductors/files/manuals/powermos6_0.pdf. Accessed 10 Feb 2018

Potnuru D, Chandra KPB, Arasaratnam I, Gu D-W, Mary KA, Ch SB (2016) Derivative-free square-root cubature kalman filter for non-linear brushless dc motors. IET Electr Power Appl 10(5):419–429

Chapter 6
Variants of Cubature Kalman Filter

6.1 Introductory Remarks

Cubature Kalman filter (CKF) discussed in the last chapter deals with nonlinear systems with single set of sensors and with Gaussian noise. In this chapter, variants of CKF, namely the cubature information filter (CIF), cubature $\mathcal{H}_\infty$ filter (C$\mathcal{H}_\infty$F) and cubature $\mathcal{H}_\infty$ information filter (C$\mathcal{H}_\infty$IF), and their square-root versions, will be explored. Each of these filters is suitable for particular applications. For example, the CIF is suitable for state estimation of nonlinear systems with multiple sensors in the presence of Gaussian noise; the C$\mathcal{H}_\infty$F is suitable for nonlinear systems with Gaussian or non-Gaussian noises; and finally, the C$\mathcal{H}_\infty$IF is useful for estimating the states of nonlinear systems with multiple sensors in the presence of Gaussian or non-Gaussian noise.

6.2 Cubature Information Filters

Similar to the Kalman filter, an information filter (IF) is also formed by two stages—prediction and update. A Kalman information filter is an algebraic equivalent of a Kalman filter where the information vector and information matrices (inverse of covariance matrices) are propagated instead of mean and covariances in the Kalman filter algorithm. Initialisation in the information space is easier (or, more tolerant) than in the Kalman filter case and the update stage of an information filter is computationally simpler than that in the Kalman filter. The information filter for linear systems and the extended information filter for nonlinear systems were introduced earlier in Chaps. 3 and 4, respectively. For the sake of convenience, they are repeated here with only the main ingredients of each of them, before the cubature information filter is discussed. One may see Chandra et al. (2013) for more details.

K. P. B. Chandra and D.-W. Gu, *Nonlinear Filtering*,
https://doi.org/10.1007/978-3-030-01797-2_6

6.2.1 Information Filters

Consider the discrete, linear process and measurement models depicted in the following

$$\mathbf{x}_k = \mathbf{F}_{k-1}\mathbf{x}_{k-1} + \mathbf{G}_{k-1}\mathbf{u}_{k-1} + \mathbf{w}_{k-1}, \tag{6.1}$$

$$\mathbf{z}_k = \mathbf{H}_k\mathbf{x}_k + \mathbf{v}_k, \tag{6.2}$$

where $k \geq 1$ is the time index, $\mathbf{x}_k$ the state vector, $\mathbf{u}_k$ the control input, $\mathbf{z}_k$ the measurement, $\mathbf{w}_{k-1}$ and $\mathbf{v}_k$ are the process and measurement noises, respectively.

The predicted information state vector, $\hat{\mathbf{y}}_{k|k-1}$, and the predicted information matrix, $\mathbf{Y}_{k|k-1}$, are defined as

$$\hat{\mathbf{y}}_{k|k-1} = \mathbf{Y}_{k|k-1}\hat{\mathbf{x}}_{k|k-1} \tag{6.3}$$

$$\mathbf{Y}_{k|k-1} = \mathbf{P}^{-1}_{k|k-1} = \left[\mathbf{F}_{k-1}\mathbf{Y}^{-1}_{k-1|k-1}\mathbf{F}^T_{k-1} + \mathbf{Q}_{k-1}\right]^{-1} \tag{6.4}$$

where $\mathbf{P}_{k|k-1}$ is the predicted covariance matrix and

$$\hat{\mathbf{x}}_{k|k-1} = \mathbf{F}_{k-1}\mathbf{x}_{k-1|k-1} + \mathbf{G}_{k-1}\mathbf{u}_{k-1}. \tag{6.5}$$

The updated information state vector, $\hat{\mathbf{y}}_{k|k}$, and the updated information matrix, $\mathbf{Y}_{k|k}$, are

$$\hat{\mathbf{y}}_{k|k} = \hat{\mathbf{y}}_{k|k-1} + \mathbf{i}_k \tag{6.6}$$

$$\mathbf{Y}_{k|k} = \mathbf{Y}_{k|k-1} + \mathbf{I}_k. \tag{6.7}$$

The information state contribution, $\mathbf{i}_k$, and its associated information matrix, $\mathbf{I}_k$, are

$$\mathbf{i}_k = \mathbf{H}^T_k\mathbf{R}^{-1}_k\mathbf{z}_k \tag{6.8}$$

$$\mathbf{I}_k = \mathbf{H}^T_k\mathbf{R}^{-1}_k\mathbf{H}_k \tag{6.9}$$

Recovery of state vector and covariance matrix is required at different stages and can be conducted, for example, by using MATLAB's *leftdivide* operator Eustice et al. (2006),

$$\hat{\mathbf{x}}_{k|k} = \mathbf{Y}_{k|k}\backslash\hat{\mathbf{y}}_{k|k} \tag{6.10}$$

$$\mathbf{P}_{k|k} = \mathbf{Y}_{k|k}\backslash\mathbf{I}_n \tag{6.11}$$

where $\mathbf{I}_n$ is the identity matrix of dimension n, with n being the size of the state vector.

6.2.2 Extended Information Filter

The extended information filter was introduced in Chap. 4. For the convenience of reference, the main formulae are repeated here. For detailed formulations and derivations of these filtering algorithms, please see Chap. 4 or Mutambara (1998) for example.

Similar to the information filter, the extended information filter is interested to find out the value of the information states and the value of the information matrix which is the inverse of the covariance matrix. However, EIF can be applied for nonlinear systems, that is not possible for the (standard) information filter. The EIF equations are to be listed below.

Consider the discrete-time, nonlinear process and measurement models as

$$\mathbf{x}_k = \mathbf{f}(\mathbf{x}_{k-1}, \mathbf{u}_{k-1}) + \mathbf{w}_{k-1} \tag{6.12}$$

$$\mathbf{z}_k = \mathbf{h}(\mathbf{x}_k, \mathbf{u}_k) + \mathbf{v}_k \tag{6.13}$$

where k is the time index, $\mathbf{x}_k$ the state vector, $\mathbf{u}_k$ the control input, $\mathbf{z}_k$ the measurement, $\mathbf{w}_{k-1}$ and $\mathbf{v}_k$ are the process and measurement noises, respectively. These noises are assumed to be zero mean, Gaussian-distributed, random variables with covariances of $\mathbf{Q}_{k-1}$ and $\mathbf{R}_k$, respectively.

The predicted information state vector, $\hat{\mathbf{y}}_{k|k-1}$, and the predicted information matrix, $\mathbf{Y}_{k|k-1}$, can be calculated by

$$\hat{\mathbf{y}}_{k|k-1} = \mathbf{Y}_{k|k-1}\hat{\mathbf{x}}_{k|k-1} \tag{6.14}$$

$$\mathbf{Y}_{k|k-1} = \mathbf{P}_{k|k-1}^{-1} = \left[\nabla\mathbf{f}_x \mathbf{Y}_{k-1|k-1}^{-1} \nabla\mathbf{f}_x^T + \mathbf{Q}_{k-1}\right]^{-1} \tag{6.15}$$

where $\mathbf{P}_{k|k-1}$ is the predicted covariance matrix and

$$\hat{\mathbf{x}}_{k|k-1} = \mathbf{f}(\hat{\mathbf{x}}_{k-1|k-1}, \mathbf{u}_{k-1}). \tag{6.16}$$

The updated information state vector, $\hat{\mathbf{y}}_{k|k}$, and the updated information matrix, $\mathbf{Y}_{k|k}$, are

$$\hat{\mathbf{y}}_{k|k} = \hat{\mathbf{y}}_{k|k-1} + \mathbf{i}_k \tag{6.17}$$

$$\mathbf{Y}_{k|k} = \mathbf{Y}_{k|k-1} + \mathbf{I}_k. \tag{6.18}$$

The information state contribution, $\mathbf{i}_k$, and its associated information matrix contribution, $\mathbf{I}_k$, are

$$\mathbf{i}_k = \nabla\mathbf{h}_x^T \mathbf{R}_k^{-1}\left[\nu_k + \nabla\mathbf{h}_x \hat{\mathbf{x}}_{k|k-1}\right] \tag{6.19}$$

$$\mathbf{I}_k = \nabla\mathbf{h}_x^T \mathbf{R}_k^{-1} \nabla\mathbf{h}_x \tag{6.20}$$

where the measurement residual, ν_k, is

$$\nu_k = \mathbf{z}_k - \mathbf{h}(\hat{\mathbf{x}}_{k|k-1}, \mathbf{u}_k) \tag{6.21}$$

and $\nabla\mathbf{f}_x$, and $\nabla\mathbf{h}_x$ are the Jacobians of $\mathbf{f}$ and $\mathbf{h}$ evaluated at the best available state, respectively.

For the nonlinear information filter, recovery of state and covariance matrices is required at different stages. As in the information filter design for linear systems, the state vector and covariance matrix can be recovered by

$$\hat{\mathbf{x}}_{k|k} = \mathbf{Y}_{k|k} \backslash \hat{\mathbf{y}}_{k|k} \tag{6.22}$$

$$\mathbf{P}_{k|k} = \mathbf{Y}_{k|k} \backslash \mathbf{I}_n \tag{6.23}$$

where $\mathbf{I}_n$ is the identity matrix of dimension n, the dimension of the state vector.

6.2.3 Cubature Information Filter

This section presents the cubature information filter (CIF) algorithm, which explores the CKF in an EIF framework. The factorisation of the error covariance matrix, $\mathbf{P}_{k-1|k-1}$, is given by,

$$\mathbf{P}_{k-1|k-1} = \mathbf{P}^{\frac{1}{2}}_{k-1|k-1} \mathbf{P}^{\frac{T}{2}}_{k-1,k-1}$$

where $\mathbf{P}^{\frac{1}{2}}_{k-1|k-1}$ is the square root factor of $\mathbf{P}_{k-1|k-1}$, the evaluation of cubature points and propagated cubature points for the process model, as required for cubature Kalman filtering process, are shown in (6.24)–(6.25).

Let the information state vector and information matrix be given by $\mathbf{y}_{k-\hat{1}|k-1}$ and $[\mathbf{Y}_{k-1|k-1}]$, respectively, and let $\mathbf{S}_{k-1|k-1}$ be a square-root factor of $[\mathbf{Y}_{k-1|k-1}]^{-1}$.

The evaluation of the cubature points and propagated cubature points can then be given as

$$\chi_{i,k-1|k-1} = \mathbf{S}_{k-1|k-1}\xi_i + \hat{\mathbf{x}}_{k-1|k-1} \tag{6.24}$$

$$\chi^*_{i,k|k-1} = \mathbf{f}(\chi_{i,k-1|k-1}, \mathbf{u}_{k-1}) \tag{6.25}$$

where $i = 1, 2, \ldots, 2n$ and n is the size of the state vector.

From (6.58) and (5.19), and (6.57) and (5.18), it yields

$$\begin{aligned} \mathbf{Y}_{k|k-1} &= \mathbf{P}^{-1}_{k|k-1} \\ &= \left[\frac{1}{2n} \sum_{i=1}^{2n} \chi^*_{i,k|k-1} \chi^{*T}_{i,k|k-1} - \hat{\mathbf{x}}_{k|k-1} \hat{\mathbf{x}}^T_{k|k-1} + \mathbf{Q}_{k-1} \right]^{-1} \end{aligned} \tag{6.26}$$

and

$$\hat{\mathbf{y}}_{k|k-1} = \mathbf{P}^{-1}_{k|k-1} \hat{\mathbf{x}}_{k|k-1} \tag{6.27}$$

$$= \mathbf{Y}_{k|k-1}\hat{\mathbf{x}}_{k|k-1} \tag{6.28}$$

$$= \frac{1}{2n}\left[\mathbf{Y}_{k|k-1}\sum_{i=1}^{2n}\boldsymbol{\chi}^{*}_{i,k|k-1}\right]. \tag{6.29}$$

In the measurement update stage of the CIF, the first two steps involve the evaluation of propagated cubature points and the predicted measurement is given below. The propagated cubature points for the measurement model can be evaluated as

$$\boldsymbol{\chi}_{i,k|k-1} = \mathbf{S}_{k|k-1}\xi_i + \hat{\mathbf{x}}_{k|k-1} \tag{6.30}$$

$$\mathbf{z}_{i,k|k-1} = \mathbf{h}(\boldsymbol{\chi}_{i,k|k-1}, \mathbf{u}_k). \tag{6.31}$$

The predicted measurement is

$$\hat{\mathbf{z}}_{k|k-1} = \frac{1}{2n}\sum_{i=1}^{2n}\mathbf{z}_{i,k|k-1}. \tag{6.32}$$

The information state contribution and its associated information matrix contribution in (6.19) and (6.20) are explicit functions of the linearised Jacobian of the measurement model. But the CKF algorithm does not require the Jacobians for measurement update and hence it cannot be directly used in the EIF framework. However, by using the following linear error propagation property Lee (2008), Sibley et al. (2006), it is possible to embed the CKF update in the EIF framework. The linear error propagation property for the error cross covariance matrix can be approximated as Sibley et al. (2006)

$$\mathbf{P}_{xz,k|k-1} \simeq \mathbf{P}_{k|k-1}\nabla\mathbf{h}_x^T. \tag{6.33}$$

By multiplying $\mathbf{P}^{-1}_{k|k-1}$ and $\mathbf{P}_{k|k-1}$ on the RHS of (6.19) and (6.20) we get

$$\mathbf{i}_k = \mathbf{P}^{-1}_{k|k-1}\mathbf{P}_{k|k-1}\nabla\mathbf{h}_x^T\mathbf{R}_k^{-1}\left[\nu_k + \nabla\mathbf{h}_x\mathbf{P}^T_{k|k-1}\mathbf{P}^{-T}_{k|k-1}\hat{\mathbf{x}}_{k|k-1}\right] \tag{6.34}$$

$$\mathbf{I}_k = \mathbf{P}^{-1}_{k|k-1}\mathbf{P}_{k|k-1}\nabla\mathbf{h}_x^T\mathbf{R}_k^{-1}\nabla\mathbf{h}_x\mathbf{P}^T_{k|k-1}\mathbf{P}^{-T}_{k|k-1}. \tag{6.35}$$

Using (6.33) in (6.34) and (6.35) we get

$$\mathbf{i}_k = \mathbf{P}^{-1}_{k|k-1}\mathbf{P}_{xz,k|k-1}\mathbf{R}_k^{-1}\left[\nu_k + \mathbf{P}^T_{xz,k|k-1}\mathbf{P}^{-T}_{k|k-1}\hat{\mathbf{x}}_{k|k-1}\right] \tag{6.36}$$

$$\mathbf{I}_k = \mathbf{P}^{-1}_{k|k-1}\mathbf{P}_{xz,k|k-1}\mathbf{R}_k^{-1}\mathbf{P}^T_{xz,k|k-1}\mathbf{P}^{-T}_{k|k-1} \tag{6.37}$$

where

$$\mathbf{P}_{xz,k|k-1} = \frac{1}{2n}\sum_{i=1}^{2n}\boldsymbol{\chi}_{i,k|k-1}\mathbf{z}^T_{i,k|k-1} - \hat{\mathbf{x}}_{k|k-1}\hat{\mathbf{z}}^T_{k|k-1}. \tag{6.38}$$

The updated information state vector and information matrix for the CIF can be obtained by using $\mathbf{i}_k$ and $\mathbf{I}_k$ from (6.36) and (6.37) in (6.17) and (6.18).

One may note that, unlike in the case of the information filter for linear systems, zero initialisation is not possible in a cubature information filter. That is because the evaluation of cubature points requires the square root of the covariance matrix. The state vector and covariance matrix in the cubature information filtering process can be recovered by (6.22) and (6.23). The CIF algorithm is summarised in Algorithm 6.1.

6.2.4 CIF in Multi-sensor State Estimation

One of the main advantages of an information filter is its ability to deal with multi-sensor data fusion (Mutambara 1998; Bar-Shalom et al. 2004). The information from various sensors can be easily fused by simply adding the information contributions to the information matrix and information vector (Mutambara 1998; Bar-Shalom et al. 2004). In the multi-sensor state estimation, the available observations consist of measurements taken from various sensors. The prediction step for the multi-sensor state estimation is similar to that of the Kalman or (single sensor) information filters. In the measurement update step, the data from various sensors are fused together for an efficient and reliable estimation (Raol and Girija 2002).

Let the various sensors used for state estimation be given by

$$\mathbf{z}_{j,k} = \mathbf{h}_{j,k}(\mathbf{x}_k, \mathbf{u}_k) + \mathbf{v}_{j,k}; \qquad j = 1, 2, \ldots D \tag{6.39}$$

where 'D' is the number of sensors.

The CIF algorithm can be easily extended for the multi-sensor data fusion in which the basic update step of CIF is similar to EIF (Mutambara 1998). The updated information vector and information matrix for multi-sensor CIF are

$$\hat{\mathbf{y}}_{k|k} = \hat{\mathbf{y}}_{k|k-1} + \sum_{j=1}^{D} \mathbf{i}_{j,k} \tag{6.40}$$

$$\mathbf{Y}_{k|k} = \mathbf{Y}_{k|k-1} + \sum_{j=1}^{D} \mathbf{I}_{j,k}. \tag{6.41}$$

By using (6.36) and (6.37), the information contributions of multi-sensor CIF are

$$\mathbf{I}_{j,k} = \mathbf{M}^T_{j,k|k-1} \mathbf{R}^{-1}_{j,k} \mathbf{M}_{j,k|k-1} \tag{6.42}$$

$$\mathbf{i}_{j,k} = \mathbf{M}^T_{j,k|k-1} \mathbf{R}^{-1}_{j,k} [\nu_{j,k} + \mathbf{M}_{j,k|k-1} \hat{\mathbf{x}}_{k|k-1}] \tag{6.43}$$

where

$$\mathbf{M}^T_{j,k|k-1} = \mathbf{P}^{-1}_{k|k-1} \mathbf{P}_{j,xz,k|k-1}. \tag{6.44}$$

Algorithm 6.1 Cubature information filter

Prediction

1: Evaluate $\chi_{i,k-1|k-1}$, $\chi^*_{i,k|k-1}$ and $\mathbf{x}_{k|k-1}$ as

$$\begin{aligned}\chi_{i,k-1|k-1} &= \sqrt{\mathbf{Y}_{k-1|k-1}\backslash\mathbb{I}_n}\xi_i + (\mathbf{Y}_{k-1|k-1}\backslash\mathbf{y}_{k-1|k-1}),\\ \chi^*_{i,k|k-1} &= \mathbf{f}(\chi_{i,k-1|k-1}, \mathbf{u}_{k-1})\\ \mathbf{x}_{k|k-1} &= \frac{1}{2n}\sum_{i=1}^{2n}\chi^*_{i,k|k-1}\end{aligned}$$

where '\' is left-divide operator, $\mathbb{I}_n$ is the nth order identity matrix, and ξ_i is the i-th element of the following set

$$\sqrt{n}\left[\begin{pmatrix}1\\0\\\vdots\\0\end{pmatrix},\ldots,\begin{pmatrix}0\\\vdots\\0\\1\end{pmatrix},\begin{pmatrix}-1\\0\\\vdots\\0\end{pmatrix},\ldots,\begin{pmatrix}0\\\vdots\\0\\-1\end{pmatrix}\right]. \tag{6.45}$$

2: The predicted information matrix and information vector are

$$\begin{aligned}\mathbf{Y}_{k|k-1} &= \mathbf{P}^{-1}_{k|k-1}\\ \hat{\mathbf{y}}_{k|k-1} &= \mathbf{Y}_{k|k-1}\frac{1}{2n}\sum_{i=1}^{2n}\chi^*_{i,k-1|k-1}\end{aligned}$$

where

$$\mathbf{P}_{k|k-1} = \frac{1}{2n}\sum_{i=1}^{2n}\chi^*_{i,k|k-1}\chi^{*T}_{i,k|k-1} - \hat{\mathbf{x}}_{k|k-1}\hat{\mathbf{x}}^T_{k|k-1} + \mathbf{Q}_{k-1}$$

Measurement Update

1: Evaluate the information state contribution and its associated information matrix contribution as

$$\begin{aligned}\mathbf{I}_k &= \mathbf{Y}_{k|k-1}\mathbf{P}_{xz,k|k-1}\mathbf{R}_k^{-1}\mathbf{P}^T_{xz,k|k-1}\mathbf{Y}^T_{k|k-1}\\ \mathbf{i}_k &= \mathbf{Y}_{k|k-1}\mathbf{P}_{xz,k|k-1}\mathbf{R}_k^{-1}\left[\nu_k + \mathbf{P}^T_{xz,k|k-1}\mathbf{Y}^T_{k|k-1}\hat{\mathbf{x}}_{k|k-1}\right]\end{aligned}$$

where

$$\mathbf{P}_{xz,k|k-1} = \frac{1}{2n}\sum_{i=1}^{2n}\chi_{i,k|k-1}\mathbf{z}^T_{i,k|k-1} - \hat{\mathbf{x}}_{k|k-1}\hat{\mathbf{z}}^T_{k|k-1}.$$

2: The estimated information vector and information matrix of CIF are

$$\begin{aligned}\mathbf{Y}_{k|k} &= \mathbf{Y}_{k|k-1} + \mathbf{I}_k\\ \hat{\mathbf{y}}_{k|k} &= \hat{\mathbf{y}}_{k|k-1} + \mathbf{i}_k.\end{aligned}$$

The state and covariance can be recovered using (6.22) and (6.23).

6.3 Cubature $\mathcal{H}_\infty$ Filters

The $\mathcal{H}_\infty$ filter, the extended $\mathcal{H}_\infty$ filter (E$\mathcal{H}_\infty$F) and the extended information filter were introduced in Chaps. 3 and 4, respectively. Again, the concept and main steps of those filtering approaches will be briefly summarised here for the sake of convenience. Based on these filters, the extended $\mathcal{H}_\infty$ information filter (E$\mathcal{H}_\infty$IF) will be introduced in this section before the introduction of the cubature $\mathcal{H}_\infty$ filter (C$\mathcal{H}_\infty$F) and cubature $\mathcal{H}_\infty$ information filter (C$\mathcal{H}_\infty$IF).

6.3.1 $\mathcal{H}_\infty$ Filters

For a general nonlinear system, the discrete-time model can be given as

$$\mathbf{x}_k = \mathbf{f}[\mathbf{x}_{k-1}, \mathbf{u}_{k-1}] + \mathbf{w}_{k-1} \tag{6.46}$$

$$\mathbf{z}_k = \mathbf{h}[\mathbf{x}_k, \mathbf{u}_k] + \mathbf{v}_k \tag{6.47}$$

where k is the time index showing the current time, $\mathbf{x}_k \in \mathcal{R}^n$ is the state vector, $\mathbf{u}_k \in \mathcal{R}^q$ the control input, $\mathbf{z}_k \in \mathcal{R}^p$ the measurement vector, and $\mathbf{w}_{k-1}$ and $\mathbf{v}_k$ are the process and observation noises, respectively.

A solution to the $\mathcal{H}_\infty$ filter based on the game theory was given in Shen and Deng (1997). Consider the process and observation models defined in (6.46) and (6.47). The noise terms $\mathbf{w}_k$ and $\mathbf{v}_k$ may be random with possibly unknown statistics, or they may be deterministic. Also, they may have zero or non-zero means. Instead of directly estimating the states, one can estimate a linear combination of the states

$$\mathbf{z}_k = \mathbf{L}_k \mathbf{x}_k \tag{6.48}$$

By replacing $\mathbf{L}$ with the identity matrix of appropriate size, it returns to the normal scenario of estimation of the state vector.

In the game theory approach to the $\mathcal{H}_\infty$ filtering, the performance measure is given by

$$\mathbf{J}_\infty = \frac{\sum_{k=0}^{N-1} \|\mathbf{z}_k - \widehat{\mathbf{z}}_k\|^2_{\mathbf{M}_k}}{\|\mathbf{x}_0 - \widehat{\mathbf{x}}_0\|^2_{\mathbf{P}_0^{-1}} + \sum_{k=0}^{N-1}(\|\mathbf{w}_k\|^2_{\mathbf{Q}_k^{-1}} + \|\mathbf{v}_k\|^2_{\mathbf{R}_k^{-1}})} \tag{6.49}$$

where $\mathbf{P}_0, \mathbf{Q}_k, \mathbf{R}_k$, and $\mathbf{M}_k$ are symmetric, positive definite, weighing matrices chosen by the user based on the problem at hand. The norm notation used in this section is $\|e\|^2_{S_k} = e^T S_k e$.

The $\mathcal{H}_\infty$ filter is designed to minimise the state estimation error so that $\mathbf{J}_\infty$ is bounded by a prescribed threshold under the "worst" possible case of $\mathbf{w}_k$, $\mathbf{v}_k$, and $\mathbf{x}_0$

$$\sup \mathbf{J}_\infty < \gamma^2 \tag{6.50}$$

where "sup" stands for supremum, and $\gamma > 0$ is the error attenuation parameter.

Based on (6.50), the designer should find $\widehat{\mathbf{x}}_k$ so that $\mathbf{J}_\infty < \gamma^2$ holds for any possible values in $\mathbf{w}_k$, $\mathbf{v}_k$, and $\mathbf{x}_0$. The best that the designer can do is to minimise $\mathbf{J}_\infty$ under worst case scenarios, then the $\mathcal{H}_\infty$ filter can be interpreted as the following 'minimax' problem

$$\min_{\widehat{\mathbf{x}}_k} \max_{\mathbf{w}_k, \mathbf{v}_k, \mathbf{x}_0} \mathbf{J}_\infty \tag{6.51}$$

The predicted state vector and auxiliary matrix are

$$\widehat{\mathbf{x}}_{k|k-1} = \mathbf{f}(\widehat{\mathbf{x}}_{k-1|k-1}, \mathbf{u}_{k-1}) \tag{6.52}$$

$$\mathbf{P}_{k|k-1} = \nabla \mathbf{f}_x \mathbf{P}_{k-1|k-1} \nabla \mathbf{f}_x^T + \mathbf{Q}_k \tag{6.53}$$

and the inverse of the updated auxiliary matrix can be obtained as

$$\mathbf{P}_{k|k}^{-1} = \mathbf{P}_{k|k-1}^{-1} + \nabla \mathbf{h}_x^T \mathbf{R}_k^{-1} \nabla \mathbf{h}_x - \gamma^{-2} \mathbf{I}_n \tag{6.54}$$

where $\mathbf{I}_n$ denotes the identity matrix of dimension $n \times n$.

The updated state is

$$\widehat{\mathbf{x}}_{k|k} = \widehat{\mathbf{x}}_{k|k-1} + \mathbf{K}_\infty [\mathbf{z}_k - \mathbf{h}(\widehat{\mathbf{x}}_{k|k-1})] \tag{6.55}$$

where

$$\mathbf{K}_\infty = \mathbf{P}_{k|k-1} \nabla \mathbf{h}_x^T [\nabla \mathbf{h}_x \mathbf{P}_{k|k-1} \nabla \mathbf{h}_x^T + \mathbf{R}_k]^{-1} \tag{6.56}$$

The Jacobians of $\mathbf{f}$ and $\mathbf{h}$, $\nabla \mathbf{f}_x$ and $\nabla \mathbf{h}_x$, are evaluated at $\widehat{\mathbf{x}}_{k-1|k-1}$ and $\widehat{\mathbf{x}}_{k|k-1}$, respectively.

6.3.2 Extended $\mathcal{H}_\infty$ Information Filter

It is required first to initialise the information vector and information matrix, $\mathcal{I}_{0|0}$ and $\mathfrak{s}_{0|0}$, for $k = 1$. The extended $\mathcal{H}_\infty$ information filter (E$\mathcal{H}_\infty$IF) procedure can then be described below.

Prediction Stage in the E$\mathcal{H}_\infty$IF:

The predicted information matrix (the inverse of the predicted covariance matrix) and the corresponding information vector are

$$\mathcal{I}_{k|k-1} = \left[\nabla \mathbf{f} \mathcal{I}_{k-1|k-1}^{-1} \nabla \mathbf{f}^T + \mathbf{Q}_{k-1} \right]^{-1} \tag{6.57}$$

$$\mathfrak{s}_{k|k-1} = \mathcal{I}_{k|k-1} \mathbf{x}_{k|k-1} \tag{6.58}$$

where

$$\mathbf{x}_{k|k-1} = \mathbf{f}(\mathcal{I}_{k|k-1}{}_{k-1|k-1}\backslash\mathfrak{s}_{k-1|k-1}, \mathbf{u}_{k-1}).$$

Note that the prediction step for the E$\mathcal{H}_\infty$IF is the same as that of the one given in the EIF in Sect. 4.3, except that $\mathbf{Q}_{k-1}$ and $\mathbf{R}_k$ are the weighting matrices rather than noise covariances. The update stage, where the sensor measurements are required, will however be different from the traditional E$\mathcal{H}_\infty$F or EIF given in Sects. 3.4 and 4.3.

Update Stage in the E$\mathcal{H}_\infty$IF:

The updated information vector and information matrix are

$$\mathfrak{s}_{k|k} = \mathfrak{s}_{k|k-1} + \mathbf{i}_k \tag{6.59}$$

$$\mathcal{I}_{k|k} = \mathcal{I}_{k|k-1} + \mathbf{I}_k. \tag{6.60}$$

where

$$\mathbf{i}_k = \nabla\mathbf{h}^T\mathbf{R}_k^{-1}\left[\nu_k + \nabla\mathbf{h}\mathbf{x}_{k|k-1}\right] \tag{6.61}$$

$$\mathbf{I}_k = \nabla\mathbf{h}^T\mathbf{R}_k^{-1}\nabla\mathbf{h} - \gamma^{-2}\mathbb{I}_n \tag{6.62}$$

and

$$\nu_k = \mathbf{z}_k - \mathbf{h}(\mathbf{x}_{k|k-1}, \mathbf{u}_k). \tag{6.63}$$

Note that the main difference between the update stages of the EIF and the E$\mathcal{H}_\infty$IF is in the information matrix contribution $\mathbf{I}_k$.

For multi-sensor state estimations, the measurement comes from various sensors and can be fused to estimate the state vector efficiently in the following way:

$$\mathfrak{s}_{k|k} = \mathfrak{s}_{k|k-1} + \sum_{j=1}^{D}\mathbf{i}_{j,k} \tag{6.64}$$

$$\mathcal{I}_{k|k} = \mathcal{I}_{k|k-1} + \sum_{j=1}^{D}\mathbf{I}_{j,k}, \tag{6.65}$$

where

$$\mathbf{i}_{j,k} = \nabla\mathbf{h}^T_{j,k|k-1}\mathbf{R}^{-1}_{j,k}[\nu_{j,k} + \nabla\mathbf{h}^T_{j,k|k-1}\mathbf{x}_{k|k-1}] \tag{6.66}$$

$$\mathbf{I}_{j,k} = \nabla\mathbf{h}^T_{j,k|k-1}\mathbf{R}^{-1}_{j,k}\nabla\mathbf{h}^T_{j,k|k-1} - \gamma^{-2}\mathbb{I}_n \tag{6.67}$$

and $\nabla\mathbf{h}^T_{j,k|k-1}$ represents the Jacobian of the nonlinear measurement function $\mathbf{h}_{j,k|k-1}$ of the 'j^{th}' sensor channel, and $\mathbf{R}_{j,k}$ is the corresponding noise.

6.3.3 Cubature $\mathcal{H}_\infty$ Filter

It can be seen from Sect. 3.4 that E$\mathcal{H}_\infty$F requires Jacobians, and the CKF discussed in Sect. 5.2 assumes that the statistical noise properties are known during the state estimation. In this subsection, we present a nonlinear estimation algorithm, the cubature $\mathcal{H}_\infty$ filter (C$\mathcal{H}_\infty$F), in which neither the Jacobians nor the statistical noise properties are used a priori for the state estimation. This approach can be useful in the presence of parametric uncertainties, too. Furthermore, if some of the statistical properties of noises are known a priori, then it can be incorporated in the proposed method. In the cubature $\mathcal{H}_\infty$ filter algorithm, the prediction step is similar to that of CKF, which involves the factorisation of the error auxiliary matrix, evaluation of cubature and propagated cubature points for the process model, and estimation of the predicted state and predicted error auxiliary matrix. Please see Arasaratnam and Haykin (2009) for more details. The update step of the E$\mathcal{H}_\infty$F requires Jacobian of the linearised measurement model, while it is not explicitly used in the CKF framework. However, by using the following linear propagation property Sibley et al. (2006), Lee (2008), Li and Jia (2010), it is possible to embed the E$\mathcal{H}_\infty$F in the CKF framework. The linear error propagation property for the error covariance and cross covariance can be approximated by Sibley et al. (2006). The reference Chandra (2014) contains more details.

$$\mathbf{P}_{xz,k|k-1} \simeq \mathbf{P}_{k|k-1}\nabla\mathbf{h}_x^T \tag{6.68}$$
$$\mathbf{P}_{zz,k|k-1} \simeq \nabla\mathbf{h}_x\mathbf{P}_{k|k-1}\nabla\mathbf{h}_x^T \tag{6.69}$$

Equation (6.68) can now be used to determine the update step of the cubature $\mathcal{H}_\infty$ filter. By multiplying $\mathbf{P}_{k|k-1}^{-1}$ and $\mathbf{P}_{k|k-1}$, and their transposes on the second term of RHS of (6.54) it yields

$$\nabla\mathbf{h}_x^T\mathbf{R}_k^{-1}\nabla\mathbf{h}_x = \mathbf{P}_{k|k-1}^{-1}\mathbf{P}_{k|k-1}\nabla\mathbf{h}_x^T\mathbf{R}_k^{-1}\nabla\mathbf{h}_x\mathbf{P}_{k|k-1}^T\mathbf{P}_{k|k-1}^{-T} \tag{6.70}$$

and by using (6.68) in (6.70) it can be shown that

$$\nabla\mathbf{h}_x^T\mathbf{R}_k^{-1}\nabla\mathbf{h}_x = \mathbf{P}_{k|k-1}^{-1}\mathbf{P}_{xz,k|k-1}\mathbf{R}_k^{-1}\mathbf{P}_{xz,k|k-1}^T\mathbf{P}_{k|k-1}^{-T} \tag{6.71}$$

By using the inverse of the updated auxiliary matrix and E$\mathcal{H}_\infty$F gain, (6.54) and (6.56), (6.68) and (6.69), the following can be obtained

$$\mathbf{K}_{c\infty} = \mathbf{P}_{xz,k|k-1}(\mathbf{P}_{zz,k|k-1} + \mathbf{R}_k)^{-1} \tag{6.72}$$
$$\mathbf{P}_{k|k}^{-1} = \mathbf{P}_{k|k-1}^{-1} + \mathbf{P}_{k|k-1}^{-1}\mathbf{P}_{xz,k|k-1}\mathbf{R}_k^{-1}\mathbf{P}_{xz,k|k-1}^T\mathbf{P}_{k|k-1}^{-T} - \gamma^{-2}\mathbf{I}_n \tag{6.73}$$

The auxiliary matrix $\mathbf{P}_{k|k}$ can be recovered from (6.73) by using MATLAB's *leftdivide* operator Eustice et al. (2006)

$$\mathbf{P}_{k|k} = \mathbf{P}_{k|k}^{-1}\backslash\mathbf{I}_n. \tag{6.74}$$

The prediction step of the cubature $\mathcal{H}_\infty$ filter is similar to the prediction step of CKF Arasaratnam and Haykin (2009). The cubature $\mathcal{H}_\infty$ filter algorithm can be therefore summarised in Algorithm 6.2.

If the statistical properties of noises are known, then in the cubature $\mathcal{H}_\infty$ filter they can be directly used as noise covariance matrices. If the statistical properties of noises are not known, then the desired performance can be achieved by using '**Q**' and '**R**' as tuning parameters. The selection of an attenuation parameter, γ, is also crucial for the filter design. In general, either an appropriate fixed γ or a time-varying γ can be used, as described in Shen and Deng (1997).

Algorithm 6.2 Cubature $\mathcal{H}_\infty$ filter

Prediction

1: Initialise the state vector, $\widehat{\mathbf{x}}$, and the auxiliary matrix, $\mathbf{P}$.
2: Factorise

$$\mathbf{P}_{k-1|k-1} = \mathbf{P}^{1/2}_{k-1|k-1}\mathbf{P}^{T/2}_{k-1,k-1}. \tag{6.75}$$

3: Evaluate the cubature points, $\boldsymbol{X}_{i,k-1|k-1}$
4: Evaluate the propagated cubature points, $\boldsymbol{X}^*_{i,k|k-1}$
5: Estimate the predicted state, $\widehat{\mathbf{x}}_{k|k-1}$
6: Estimate the predicted auxiliary matrix, $\mathbf{P}_{k|k-1}$

Measurement update

1: Factorise

$$\mathbf{P}_{k|k-1} = \mathbf{P}^{1/2}_{k|k-1}\mathbf{P}^{T/2}_{k|k-1}. \tag{6.76}$$

2: Evaluate the cubature points

$$\boldsymbol{X}_{i,k|k-1} = \mathbf{P}^{1/2}_{k|k-1}\xi_i + \widehat{\mathbf{x}}_{k|k-1}. \tag{6.77}$$

3: Evaluate the propagated cubature points of measurement model, $\mathbf{Z}_{i,k|k-1}$
4: Estimate the predicted measurement, $\widehat{\mathbf{z}}_{k|k-1}$
5: Estimate the measurement auxiliary matrix, $\mathbf{P}_{zz,k|k-1}$
6: Estimate the cross-auxiliary matrix, $\mathbf{P}_{xz,k|k-1}$
7: Estimate the gain matrix, $\mathbf{K}_{c\infty}$, using (6.72)
8: Estimate the updated state

$$\widehat{\mathbf{x}}_{k|k} = \widehat{\mathbf{x}}_{k|k-1} + \mathbf{K}_{c\infty}(\mathbf{z}_k - \widehat{\mathbf{z}}^T_{k|k-1}). \tag{6.78}$$

9: Estimate the inverse of the updated auxiliary matrix, $\mathbf{P}^{-1}_{k|k}$, by using (6.73).
10: Recover the auxiliary matrix, $\mathbf{P}_{k|k}$, using (6.74).

6.3.4 Cubature $\mathcal{H}_\infty$ Information Filter

This subsection presents the derivation of the cubature $\mathcal{H}_\infty$ information filter (C$\mathcal{H}_\infty$IF), and its extension to be applied in the multi-sensor state estimation. The key idea is to fuse the CKF and E$\mathcal{H}_\infty$IF to form a derivative-free, state estimator for nonlinear systems. The E$\mathcal{H}_\infty$IF can deal with multi-sensor scenario for Gaussian and non-Gaussian noises, but it needs the Jacobians which would compromise the accuracy of state estimation. The linear error propagation property will be used to derive the C$\mathcal{H}_\infty$IF. The derived filter will have the desired properties of the CKF and of the E$\mathcal{H}_\infty$IF including derivative-free estimation, the ability to handle Gaussian and non-Gaussian noises as well as multi-sensor state estimation. The prediction stage in the C$\mathcal{H}_\infty$IF will be the same as that of the CKF already given in Sect. 5.2. Interested readers may find more details in Chandra (2016). However, the cubature $\mathcal{H}_\infty$ information filtering process will propagate the information vector and the information matrix as will be shown in the prediction stage in the algorithm below.

The updated information vector and information matrix are given by

$$\mathfrak{s}_{k|k} = \mathfrak{s}_{k|k-1} + \mathbf{i}_k, \tag{6.79}$$

$$\mathcal{I}_{k|k} = \mathcal{I}_{k|k-1} + \mathbf{I}_k. \tag{6.80}$$

The update stage for the E$\mathcal{H}_\infty$IF given in Sect. 6.3.2 will be explored to develop the update stage for the C$\mathcal{H}_\infty$IF. Consider the linear error propagation property (Chandra 2014; Lee 2008; Sibley et al. 2006; Li and Jia 2010)

$$\mathbf{P}_{zz,k|k-1} \simeq \nabla\mathbf{h}\mathbf{P}_{k|k-1}\nabla\mathbf{h}^T, \tag{6.81}$$

$$\mathbf{P}_{xz,k|k-1} \simeq \mathbf{P}_{k|k-1}\nabla\mathbf{h}^T. \tag{6.82}$$

Pre-multiplying by $\mathbf{P}_{k|k-1}^{-1}$ on both sides of (6.82) yields

$$\begin{aligned}\nabla\mathbf{h}^T &= \mathbf{P}_{k|k-1}^{-1}\mathbf{P}_{xz,k|k-1}, \\ &= \mathcal{I}_{k|k-1}\mathbf{P}_{xz,k|k-1}.\end{aligned} \tag{6.83}$$

Using (6.83) in (6.61) and (6.62), the following can be found:

$$\mathbf{i}_k = \mathcal{I}_{k|k-1}\mathbf{P}_{xz,k|k-1}\mathbf{R}_k^{-1}[\nu_k + \mathbf{P}_{xz,k|k-1}^T\mathcal{I}_{k|k-1}^T\mathbf{x}_{k|k-1}], \tag{6.84}$$

$$\mathbf{I}_k = \mathcal{I}_{k|k-1}\mathbf{P}_{xz,k|k-1}\mathbf{R}_k^{-1}\mathbf{P}_{xz,k|k-1}^T\mathcal{I}_{k|k-1}^T - \gamma^{-2}\mathbb{I}_n. \tag{6.85}$$

The derivative-free information matrix and cross error covariance matrix, $\mathcal{I}_{k|k-1} = \mathbf{P}_{k|k-1}^{-1}$ and $\mathbf{P}_{xz,k|k-1}$, can be obtained from (5.19) and (5.22). The information contributions in (6.84) and (6.85), along with (6.79) and (6.80), represent the update stage of the C$\mathcal{H}_\infty$IF.

As mentioned earlier, one of the key advantages of using information filters is their ability to estimate efficiently the states with multiple sensors. There are several applications where multiple sensors are preferred to estimate the states such as the target tracking of a re-entry vehicle, where the aircraft states are estimated using radars located at different altitudes Lee (2008), etc. The update step described above for the $C\mathcal{H}_\infty$IF can easily be extended for the case of multiple sensors. The structure of the update stage for the multi-sensor $C\mathcal{H}_\infty$IF will be the same as that of the $E\mathcal{H}_\infty$IF given in Sect. 6.3.2, but the information contribution factors $\mathbf{i}_{j,k}$ and $\mathbf{I}_{j,k}$ will be different, and they are derivative-free. The updated information vector and the corresponding information matrix contribution for $C\mathcal{H}_\infty$IF with multiple sensors are

$$\mathfrak{s}_{k|k} = \mathfrak{s}_{k|k-1} + \sum_{j=1}^{D} \mathbf{i}_{j,k} \tag{6.86}$$

$$\mathcal{I}_{k|k} = \mathcal{I}_{k|k-1} + \sum_{j=1}^{D} \mathbf{I}_{j,k}, \tag{6.87}$$

where

$$\mathbf{i}_{j,k} = \mathcal{I}_{j,k|k-1}\mathbf{P}_{xz,j,k|k-1}\mathbf{R}_{j,k}^{-1}[\nu_{j,k} + \mathbf{P}^T_{xz,j,k|k-1}\mathcal{I}^T_{j,k|k-1}\mathbf{x}_{k|k-1}] \tag{6.88}$$

$$\mathbf{I}_{j,k} = \mathcal{I}_{j,k|k-1}\mathbf{P}_{xz,j,k|k-1}\mathbf{R}_{j,k}^{-1}\mathbf{P}^T_{xz,j,k|k-1}\mathcal{I}^T_{j,k|k-1} - \gamma^{-2}\mathbb{I}_n \tag{6.89}$$

The $C\mathcal{H}_\infty$IF and its extension to the case of multiple sensors are summarised in the algorithm next.

Algorithm 6.3 Cubature $\mathcal{H}_\infty$ information filter

Initialise the information vector and information matrix, $\mathcal{I}_{0|0}$ and $\mathfrak{s}_{0|0}$ for $k = 1$.

Prediction

1: Evaluate $\chi_{i,k-1|k-1}$, $\chi^*_{i,k|k-1}$ and $\mathbf{x}_{k|k-1}$ as

$$\chi_{i,k-1|k-1} = \sqrt{\mathcal{I}_{k-1|k-1}\backslash\mathbb{I}_n}\xi_i + (\mathcal{I}_{k-1|k-1}\backslash\mathfrak{s}_{k-1|k-1}),$$

$$\chi^*_{i,k|k-1} = \mathbf{f}(\chi_{i,k-1|k-1}, \mathbf{u}_{k-1})$$

$$\mathbf{x}_{k|k-1} = \frac{1}{2n}\sum_{i=1}^{2n}\chi^*_{i,k|k-1}$$

where '\' is left-divide operator, $\mathbb{I}_n$ is the nth order identity matrix, and where ξ_i is the i-th element of the following set:

$$\sqrt{n}\left[\begin{pmatrix}1\\0\\\vdots\\0\end{pmatrix},\ldots,\begin{pmatrix}0\\\vdots\\0\\1\end{pmatrix},\begin{pmatrix}-1\\0\\\vdots\\0\end{pmatrix},\ldots,\begin{pmatrix}0\\\vdots\\0\\-1\end{pmatrix}\right] \tag{6.90}$$

2: The predicted information matrix and information vector are

$$\mathcal{I}_{k|k-1} = \left[\frac{1}{2n}\sum_{i=1}^{2n} \chi^*_{i,k|k-1}\chi^{*T}_{i,k|k-1} - \mathbf{x}_{k|k-1}\mathbf{x}^T_{k|k-1} + \mathbf{Q}_{k-1}\right] \backslash \mathbb{I}_n$$

$$\mathfrak{s}_{k|k-1} = \frac{\mathcal{I}_{k|k-1}}{2n}\left[\sum_{i=1}^{2n} \chi^*_{i,k-1|k-1}\right].$$

Measurement update

1: Evaluate $\chi_{i,k|k-1}$, $\mathbb{Z}_{i,k|k-1}$ and $\mathbf{z}_{k|k-1}$ as

$$\chi_{i,k|k-1} = \sqrt{\mathcal{I}_{k|k-1}\backslash\mathbb{I}_n}\xi_i + \mathbf{x}_{k|k-1}$$

$$\mathbb{Z}_{i,k|k-1} = \mathbf{h}(\chi_{i,k|k-1}, \mathbf{u}_k)$$

$$\mathbf{z}_{k|k-1} = \frac{1}{2n}\sum_{i=1}^{2n} \mathbb{Z}_{i,k|k-1}$$

2: Evaluate

$$\mathbf{P}_{xz,k|k-1} = \frac{1}{2n}\sum_{i=1}^{2n} \chi_{i,k|k-1}\mathbb{Z}^T_{i,k|k-1} - \mathbf{x}_{k|k-1}\mathbf{z}^T_{k|k-1}$$

3: The information contributions are

$$\mathbf{i}_k = \mathcal{I}_{k|k-1}\mathbf{P}_{xz,k|k-1}\mathbf{R}_k^{-1}[\nu_k + \mathbf{P}^T_{xz,k|k-1}\mathcal{I}^T_{k|k-1}\mathbf{x}_{k|k-1}]$$

$$\mathbf{I}_k = \mathcal{I}_{k|k-1}\mathbf{P}_{xz,k|k-1}\mathbf{R}_k^{-1}\mathbf{P}^T_{xz,k|k-1}\mathcal{I}^T_{k|k-1} - \gamma^{-2}\mathbb{I}_n,$$

where γ is the attenuation parameter.

4: Finally, the update information vector and information matrix are

$$\mathfrak{s}_{k|k} = \mathfrak{s}_{k|k-1} + \mathbf{i}_k,$$

$$\mathcal{I}_{k|k} = \mathcal{I}_{k|k-1} + \mathbf{I}_k$$

Measurement update for multi-sensor state estimation

$$\mathfrak{s}_{k|k} = \mathfrak{s}_{k|k-1} + \sum_{j=1}^{D} \mathbf{i}_{j,k}$$

$$\mathcal{I}_{k|k} = \mathcal{I}_{k|k-1} + \sum_{j=1}^{D} \mathbf{I}_{j,k},$$

where

$$\mathbf{I}_{j,k} = \mathcal{I}_{j,k|k-1}\mathbf{P}_{xz,j,k|k-1}\mathbf{R}_{j,k}^{-1}\mathbf{P}^T_{xz,j,k|k-1}\mathcal{I}^T_{j,k|k-1} - \gamma^{-2}\mathbb{I}_n$$

$$\mathbf{i}_{j,k} = \mathcal{I}_{j,k|k-1}\mathbf{P}_{xz,j,k|k-1}\mathbf{R}_{j,k}^{-1}[\nu_{j,k} + \mathbf{P}^T_{xz,j,k|k-1}\mathcal{I}^T_{j,k|k-1}\mathbf{x}_{k|k-1}].$$

6.4 Square-Root Version Filters

One of the most numerically stable and reliable implementations of the Kalman filter is its square-root version (Grewal and Andrews 2001). In this approach the square root of the covariance matrix is propagated to make the overall filter *robust* against round-off errors. Some of the key properties of the square-root filters include symmetric positive definite error covariances, availability of square-root factors, doubled-order precision, and therefore improved numerical accuracy (Park and Kailath 1995; Grewal and Andrews 2001; Simon 2006; Arasaratnam and Haykin 2009). Similarly, in the information domain, the square-root versions of information filters are preferred (Park and Kailath 1995). These added advantages of square-root filters are the motivation for development of square-root cubature filters. This section presents a brief description of the square-root versions of the extended Kalman filter (SREKF), the square-root extended information filter (SREIF), the square-root extended $\mathcal{H}_\infty$ filter (SRE$\mathcal{H}_\infty$F), the square-root cubature Kalman filter (SRCKF), the square-root cubature information filter (SRCIF), the square-root cubature $\mathcal{H}_\infty$ filter (SRC$\mathcal{H}_\infty$F) and the square-root cubature $\mathcal{H}_\infty$ information filter (SRC$\mathcal{H}_\infty$IF).

The following notation is used throughout this section.

Given a positive definite matrix F, then F and F^{-1} can be factorised as

$$\mathbf{F} = (\mathbf{F}^{1/2})(\mathbf{F}^{T/2}) \tag{6.91}$$

$$\mathbf{F}^{-1} = (\mathbf{F}^{-T/2})(\mathbf{F}^{-1/2}). \tag{6.92}$$

6.4.1 *Square-Root Extended Kalman Filter*

The two stages for the square-root extended Kalman filter (SREKF) are given in Algorithm 6.4 below.

By squaring both the sides of (6.94) and by doing some straight manipulations on (6.95), it can be shown that the EKF and its square-root version are equivalent (Park and Kailath 1995). However, due to the specific structure of the covariance matrix in the square-root version, it always ensures the positive definiteness of the error covariance matrix. That is obviously a numerical advantage.

Algorithm 6.4 Square-root extended Kalman filter

Initialise the state vector and the covariance $\mathbf{x}_0$ and $\mathbf{P}_{0|0}$ (set $k = 1$).

Prediction

1: Factorise the covariance matrix, $\mathbf{P}_{k-1|k-1}$

$$\mathbf{P}_{k-1|k-1} = \mathbf{P}^{\frac{1}{2}}_{k-1|k-1}\mathbf{P}^{\frac{T}{2}}_{k-1|k-1}$$

where $\mathbf{P}^{\frac{1}{2}}_{k-1|k-1}$ is the square-root factor of $\mathbf{P}_{k-1|k-1}$.

2: The predicted state is given by

$$\hat{\mathbf{x}}_{k|k-1} = \mathbf{f}(\mathbf{x}_{k-1|k-1}, \mathbf{u}_{k-1})$$

3: The predicted covariance $\mathbf{P}^{\frac{1}{2}}_{k|k-1}$ can be calculated from

$$\left[\nabla\mathbf{f}_x\mathbf{P}^{\frac{1}{2}}_{k-1|k-1} \quad \mathbf{Q}^{\frac{1}{2}}_k\right]\Lambda = \left[\mathbf{P}^{\frac{1}{2}}_{k|k-1} \quad \mathbf{0}\right], \tag{6.93}$$

where, Λ is a unitary matrix (Grewal and Andrews 2001) which can be computed by QR decomposition, $\nabla\mathbf{f}_x$ is the Jacobian of $\mathbf{f}$ evaluated at $\mathbf{x}_{k-1|k-1}$ and $\mathbf{Q}^{\frac{1}{2}}_k$ is the square root of the process noise covariance matrix. The predicted covariance $\mathbf{P}^{\frac{1}{2}}_{k|k-1}$ is evaluated based on the previous time sample covariance $\mathbf{P}^{\frac{1}{2}}_{k-1|k-1}$.

Measurement update

1: The updated covariance $\mathbf{P}^{\frac{1}{2}}_{k|k}$, the auxiliary matrix $\mathbf{R}^{\frac{1}{2}}_{e,k}$ and the Kalman gain matrix $\mathbf{K}_k$ required for updating the state can be obtained from

$$\begin{bmatrix} \mathbf{R}^{\frac{1}{2}}_k & \nabla\mathbf{h}_x\mathbf{P}^{\frac{1}{2}}_{k|k-1} \\ \mathbf{0} & \mathbf{P}^{\frac{1}{2}}_{k|k-1} \end{bmatrix}\Lambda = \begin{bmatrix} \mathbf{R}^{\frac{1}{2}}_{e,k} & \mathbf{0} \\ \mathbf{K}_k & \mathbf{P}^{\frac{1}{2}}_{k|k} \end{bmatrix}, \tag{6.94}$$

where, $\nabla\mathbf{h}_x$ is the Jacobian of $\mathbf{h}$ evaluated at $\mathbf{x}_{k|k-1}$, $\mathbf{R}^{\frac{1}{2}}_k$ is the square root of the measurement noise covariance matrix.

2: The updated state can then be obtained from

$$\mathbf{x}_{k|k} = \mathbf{x}_{k|k-1} + \mathbf{K}_k\mathbf{R}^{-\frac{1}{2}}_{e,k}[\mathbf{z}_k - \mathbf{h}(\mathbf{x}_{K|k-1})]. \tag{6.95}$$

[1]

6.4.2 Square-Root Extended Information Filter

In this subsection, the square-root extended information filter (SREIF), which is required for the square-root cubature information filter (SRCIF) derivation later, is to be briefly discussed. The prediction step of SREIF is the same as in the extended information filter (EIF) and hence it is not presented here. The measurement update step for SREIF is as follows Kailath et al. (2000):

$$\begin{bmatrix} \mathbf{P}_{k|k-1}^{-T/2} & \nabla\mathbf{h}_x^T\mathbf{R}_k^{-T/2} \\ \hat{\mathbf{x}}_{k|k-1}^T\mathbf{P}_{k|k-1}^{-T/2} & \mathbf{z}_k^T\mathbf{R}_k^{-T/2} \end{bmatrix}\Theta = \begin{bmatrix} \mathbf{P}_{k|k}^{-T/2} & \mathbf{0} \\ \hat{\mathbf{x}}_{k|k}^T\mathbf{P}_{k|k}^{-T/2} & \star \end{bmatrix} \tag{6.96}$$

and in the information space, it is

$$\begin{bmatrix} \mathbf{Y}_{S,k|k-1} & \nabla\mathbf{h}_x^T\mathbf{Y}_{\mathbf{R}} \\ \hat{\mathbf{y}}_{S,k|k-1}^T & \mathbf{z}_k^T\mathbf{Y}_{\mathbf{R}} \end{bmatrix}\Theta = \begin{bmatrix} \mathbf{Y}_{S,k|k} & \mathbf{0} \\ \hat{\mathbf{y}}_{S,k|k}^T & \star \end{bmatrix} \tag{6.97}$$

where $\hat{\mathbf{y}}_S$, $\mathbf{Y}_S$ and $\mathbf{Y}_{\mathbf{R}}$ are the square-root factors[1] of $\mathbf{y}$, $\mathbf{Y}$ and $\mathbf{R}$, respectively. '$\star$' represents the terms which are irrelevant for SREIF and Θ is a unitary matrix which can be found using Givens rotations or Householder reflections[2] Grewal and Andrews (2001). If Θ is partitioned as $[\Theta_1 \quad \Theta_2; \Theta_3 \quad \Theta_4]$, then the updated information state vector and the corresponding information matrix can be written as

$$\hat{\mathbf{y}}_{S,k|k}^T = \mathbf{y}_{S,k|k-1}^T\Theta_1 + \mathbf{z}_k^T\mathbf{Y}_{\mathbf{R}}\Theta_3 \tag{6.98}$$

$$\mathbf{Y}_{S,k|k} = \mathbf{Y}_{S,k|k-1}\Theta_1 + \nabla\mathbf{h}_x^T\mathbf{Y}_{\mathbf{R}}\Theta_3. \tag{6.99}$$

The measurement update stage of SREIF can be extended for multi-sensor state estimation Raol and Girija (2002). The data from different sensors are fused for an efficient and reliable estimation.

The measurement update step for SREIF using 'D' sensors is given in (6.100).

$$\begin{bmatrix} \mathbf{Y}_{S,k|k-1} & \nabla\mathbf{h}_{1,k}^T\mathbf{Y}_{\mathbf{R},1,k} & \nabla\mathbf{h}_{2,k}^T\mathbf{Y}_{\mathbf{R},2,k} & \dots & \nabla\mathbf{h}_{D,k}^T\mathbf{Y}_{\mathbf{R},D,k} \\ \hat{\mathbf{y}}_{S,k|k-1}^T & \mathbf{z}_{1,k}^T\mathbf{Y}_{\mathbf{R},1,k} & \mathbf{z}_{2,k}^T\mathbf{Y}_{\mathbf{R},2,k} & \dots & \mathbf{z}_{D,k}^T\mathbf{Y}_{\mathbf{R},D,k} \end{bmatrix}\Theta = \begin{bmatrix} \mathbf{Y}_{S,k|k} & \mathbf{0} \\ \hat{\mathbf{y}}_{S,k|k}^T & \star \end{bmatrix}. \tag{6.100}$$

6.4.3 Square-Root Extended $\mathcal{H}_\infty$ Filter

Along the lines of the square-root Kalman filters, a few researchers have explored the square-root $\mathcal{H}_\infty$ filters. In Hassibi et al. (2000), the square-root algorithms for

[1] Square-root factors of the information matrix and information state

$$\begin{aligned} \mathbf{P}^{-1} &= \mathbf{P}^{-T/2}\mathbf{P}^{-1/2} \\ \Rightarrow \mathbf{Y} &= \mathbf{Y}_s\mathbf{Y}_s^T \\ \mathbf{P}^{-1}\mathbf{x} &= \mathbf{P}^{-T/2}\mathbf{P}^{-1/2}\mathbf{x} \\ \Rightarrow \mathbf{y} &= \mathbf{Y}_s\mathbf{y}_s. \end{aligned}$$

[2] The basic structure of the Householder matrix, Θ, is

$$\Theta = \mathbf{I} - \frac{2}{c^Tc}cc^T$$

where c is a column vector and $\mathbf{I}$ is the identity matrix of the same dimension.

$\mathcal{H}_\infty$ a priori, a posteriori and filtering problems are developed in the Krein space. The square-root $\mathcal{H}_\infty$ information estimation for a rectangular discrete-time descriptor system is described in Terra (2009), where the inverse of the covariance matrices (information matrices) are propagated. In Ishihara (2006), the square-root $\mathcal{H}_\infty$ estimators for general descriptor systems are developed. One of the major differences in deriving the square-root Kalman filter and square-root $\mathcal{H}_\infty$ filter is the use of a rotation matrix. The square-root Kalman filter uses the unitary matrix, whereas the square-root $\mathcal{H}_\infty$ filter uses the **J**-unitary matrix.

In this subsection, we will derive the update step of the square-root extended $\mathcal{H}_\infty$ filter. The square-root cubature $\mathcal{H}_\infty$ filter will be then discussed in a later subsection. The following notations for matrices are used in this and later subsections.

Given a symmetric matrix $\mathbf{F}$, then $\mathbf{F}$ and $\mathbf{F}^{-1}$ can be factorised as

$$\mathbf{F} = (\mathbf{F}^{1/2})S(\mathbf{F}^{T/2}) \tag{6.101}$$

$$\mathbf{F}^{-1} = (\mathbf{F}^{-T/2})S(\mathbf{F}^{-1/2}) \tag{6.102}$$

where S is the signature matrix. When $\mathbf{F}$ is nonnegative definite, then the S is the identity matrix and $\mathbf{F}^{1/2}$ is a conventional square-root factor of $\mathbf{F}$.

The measurement update step of the square-root extended $\mathcal{H}_\infty$ filter is

$$\begin{bmatrix} -\nabla\mathbf{h}_x^T\mathbf{R}_k^{-T/2} & \mathbf{P}_{k|k-1}^{-T/2} & \gamma^{-1}\mathbf{I}_n \\ \mathbf{R}_k^{1/2} & \nabla\mathbf{h}_x\mathbf{P}_{k|k-1}^{1/2} & \mathbf{0} \end{bmatrix} \Theta_J = \begin{bmatrix} \mathbf{0} & \mathbf{P}_{k|k}^{-T/2} & \mathbf{0} \\ \mathbf{R}_e^{1/2} & \mathbf{0} & \mathbf{0} \end{bmatrix} \tag{6.103}$$

where Θ_J is a **J**-unitary matrix.

Once the square-root factors of the auxiliary matrices $\mathbf{P}_{k|k}$ and $\mathbf{R}_e = \nabla\mathbf{h}_x\mathbf{P}_{k|k-1}\nabla\mathbf{h}_x^T + \mathbf{R}_k$ are obtained, the gain matrix of square-root extended $\mathcal{H}_\infty$ filter can be obtained as

$$\mathbf{K}_{S\infty,k} = \mathbf{P}_{k|k-1}^{1/2}\mathbf{P}_{k|k-1}^{T/2}\nabla\mathbf{h}_x^T\mathbf{R}_e^{-T/2}\mathbf{R}_e^{-1/2} \tag{6.104}$$

The update state of square-root extended $\mathcal{H}_\infty$ filter is

$$\widehat{\mathbf{x}}_{k|k} = \widehat{\mathbf{x}}_{k|k-1} + \mathbf{K}_{S\infty,k}(\mathbf{z}_k - \mathbf{h}(\widehat{\mathbf{x}}_{k|k-1})) \tag{6.105}$$

The measurement update in (6.103) can be easily derived by squaring both sides,[3]

$$\begin{bmatrix} -\nabla\mathbf{h}_x^T\mathbf{R}_k^{-T/2} & \mathbf{P}_{k|k-1}^{-T/2} & \gamma^{-1}\mathbf{I}_n \\ \mathbf{R}_k^{1/2} & \nabla\mathbf{h}_x\mathbf{P}_{k|k-1}^{1/2} & \mathbf{0} \end{bmatrix} \Theta_J\mathbf{J}\Theta_J^T \begin{bmatrix} -\mathbf{R}_k^{-1/2}\nabla\mathbf{h}_x & \mathbf{R}_k^{T/2} \\ \mathbf{P}_{k|k-1}^{-1/2} & \mathbf{P}_{k|k-1}^{T/2}\nabla\mathbf{h}_x^T \\ \gamma^{-1}\mathbf{I}_n & \mathbf{0} \end{bmatrix}$$

---

[3]If $b = a\Theta$, with Θ as J-unitary matrix, then Hassibi et al. (2000)

$$bJb^T = a\Theta J\Theta^T a^T = aJa^T.$$

$$= \begin{bmatrix} \mathbf{0} & \mathbf{P}_{k|k}^{-T/2} & \mathbf{0} \\ \mathbf{R}_e^{1/2} & \mathbf{0} & \mathbf{0} \end{bmatrix} \mathbf{J} \begin{bmatrix} \mathbf{0} & \mathbf{R}_e^{T/2} \\ \mathbf{P}_{k|k}^{-1/2} & \mathbf{0} \\ \mathbf{0} & \mathbf{0} \end{bmatrix} \tag{6.106}$$

The **J**-unitary matrix for (6.106) can be chosen as

$$\mathbf{J} = \begin{bmatrix} \mathbf{I}_p & \mathbf{0} & \mathbf{0} \\ \mathbf{0} & \mathbf{I}_n & \mathbf{0} \\ \mathbf{0} & \mathbf{0} & -\mathbf{I}_n \end{bmatrix} \tag{6.107}$$

where $\mathbf{I}_p$ and $\mathbf{I}_n$ denotes the identity matrices of dimension $p \times p$ and $n \times n$, respectively.

By using $\Theta_J \mathbf{J} \Theta_J^T = \mathbf{J}$, (6.106) can further be written as

$$\begin{bmatrix} \mathbf{P}_{k|k-1}^{-1} + \nabla\mathbf{h}_x^T \mathbf{R}_k^{-1} \nabla\mathbf{h}_x - \gamma^{-2}\mathbf{I}_n & \mathbf{0} \\ \mathbf{0} & \nabla\mathbf{h}_x \mathbf{P}_{k|k-1} \nabla\mathbf{h}_x^T + \mathbf{R}_k \end{bmatrix} = \begin{bmatrix} \mathbf{P}_{k|k}^{-1} & \mathbf{0} \\ \mathbf{0} & \mathbf{R}_e \end{bmatrix} \tag{6.108}$$

It can be seen that the first entries of (6.108) and of (6.104) are the same as the inverse of the updated auxiliary matrix and the gain of the $\mathcal{H}_\infty$ filter described in Sect. 3.4, and hence the square-root extended $\mathcal{H}_\infty$ filter derived in this subsection is equivalent to the $\mathcal{H}_\infty$ filter in Sect. 3.4.

6.4.4 Square-Root Cubature Kalman Filter

Similar to the square-root EKF, the square-root CKF also propagates a square root of the predicted and updated covariance matrices. The main steps required for the square-root CKF is to be briefed in this subsection, for more details see Arasaratnam and Haykin (2009). The square-root CKF can also be expressed in two stages, prediction and update stages. The first step for the prediction is the initialisation of the state vector and the square-root error covariance matrix, $\hat{\mathbf{x}}_{0|0}$ and $\sqrt{\mathbf{P}_{0|0}}$. The predicted state can be computed using the below procedure. First, calculate the cubature points

$$\chi_{i,k-1|k-1} = \mathbf{P}_{k-1|k-1}^{\frac{1}{2}} \xi_i + \hat{\mathbf{x}}_{k-1|k-1}, \quad i = 1, \ldots, 2n$$

where ξ_i is the ith element of the following set:

$$\sqrt{n} \left[\begin{pmatrix} 1 \\ 0 \\ \vdots \\ 0 \end{pmatrix}, \ldots, \begin{pmatrix} 0 \\ \vdots \\ 0 \\ 1 \end{pmatrix}, \begin{pmatrix} -1 \\ 0 \\ \vdots \\ 0 \end{pmatrix}, \ldots, \begin{pmatrix} 0 \\ \vdots \\ 0 \\ -1 \end{pmatrix} \right].$$

The cubature points need to be propagated through the nonlinear process model as

$$\chi^*_{i,k-1|k-1} = \mathbf{f}(\chi_{i,k-1|k-1}, \mathbf{u}_{k-1}), \quad i = 1, \ldots, 2n.$$

Finally, the predicted state can be obtained as

$$\hat{\mathbf{x}}_{k|k-1} = \frac{1}{2n}\sum_{i=1}^{2n} \chi^*_{i,k-1|k-1}$$

The square-root factor of the predicted error covariance is

$$\sqrt{\mathbf{P}}_{k|k-1} = \left[\mathbf{qr}\left(\mathcal{X}^*_{i,k|k-1} \qquad \sqrt{\mathbf{Q}_{k-1}}\right)^T\right]^T \tag{6.109}$$

where

$$\mathcal{X}^*_{i,k|k-1} = \frac{1}{\sqrt{2n}}\Big[\chi^*_{1,k|k-1} - \hat{\mathbf{x}}_{k|k-1} \quad \chi^*_{2,k|k-1} - \hat{\mathbf{x}}_{k|k-1} \quad \ldots \quad \chi^*_{2n,k|k-1} - \hat{\mathbf{x}}_{k|k-1}\Big]. \tag{6.110}$$

Once the predicted state and square-root covariance factor of the square-root CKF is obtained, the next step is to find the updated state and updated square-root updated covariance. The predicted measurement, $\mathbf{z}_{i,k|k-1}$, is

$$\mathbf{z}_{i,k|k-1} = \mathbf{h}(\chi_{i,k|k-1}, \mathbf{u}_k) \tag{6.111}$$

$$\chi_{i,k|k-1} = \sqrt{\mathbf{P}_{k|k-1}}\xi_i + \hat{\mathbf{x}}_{k|k-1}. \tag{6.112}$$

The square-root factors of innovation covariance and cross-covariance matrices are

$$\sqrt{\mathbf{P}}_{zz,k|k-1} = \left[\mathbf{qr}\left(\mathcal{Z}_{k|k-1} \quad \sqrt{\mathbf{R}_k}\right)^T\right]^T \tag{6.113}$$

$$\mathbf{P}_{xz,k|k-1} = \mathcal{X}_{k|k-1}\mathcal{Z}^T_{k|k-1} \tag{6.114}$$

where,

$$\mathcal{Z}_{k|k-1} = \frac{1}{\sqrt{2n}}\Big[\mathbf{z}_{1,k|k-1} - \hat{\mathbf{z}}_{k|k-1} \quad \mathbf{z}_{2,k|k-1} - \hat{\mathbf{z}}_{k|k-1} \ldots \quad \mathbf{z}_{2n,k|k-1} - \hat{\mathbf{z}}_{k|k-1}\Big] \tag{6.115}$$

$$\mathcal{X}_{k|k-1} = \frac{1}{\sqrt{2n}}\Big[\mathbf{X}_{1,k|k-1} - \hat{\mathbf{x}}_{k|k-1} \quad \mathbf{X}_{2,k|k-1} - \hat{\mathbf{x}}_{k|k-1} \ldots \quad \mathbf{X}_{2n,k|k-1} - \hat{\mathbf{x}}_{k|k-1}\Big]. \tag{6.116}$$

The updated state and the updated square-root factor of the error covariance matrix can then be calculated from

$$\hat{\mathbf{x}}_{k|k} = \hat{\mathbf{x}}_{k|k-1} + \mathbf{K}_k(\mathbf{z}_k - \hat{\mathbf{z}}_{k|k-1}) \tag{6.117}$$

$$\sqrt{\mathbf{P}_{k|k}} = \left[\mathbf{qr}\left(\mathcal{X}_{k|k-1} - \mathbf{K}_k\mathcal{Z}_{k|k-1} \quad \mathbf{K}_k\sqrt{\mathbf{R}_k}\right)^T\right]^T, \tag{6.118}$$

where,

$$\mathbf{K}_k = \left[\mathbf{P}_{xz,k|k-1}/\sqrt{\mathbf{P}}^T_{zz,k|k-1}\right]/\sqrt{\mathbf{P}}_{zz,k|k-1}. \tag{6.119}$$

6.4.5 Square-Root Cubature Information Filter

In this subsection, another information filter using the square-root approach for nonlinear systems is introduced. This algorithm is derived from SREIF (Park and Kailath 1995) and CKF discussed earlier in the book, and is called as the square-root cubature information filter (SRCIF). The square-root factors of covariance matrices can be found using the **QR**[4] decomposition and the *leftdivide* operator. Furthermore, this approach can be extended to the multi-sensor data fusion, too.

The SRCIF is detailed in Algorithm 6.5. As in (6.127), the square-root factor of the predicted error covariance matrix is evaluated. Similarly, the square-root factor of the measurement error covariance matrix, $\mathbf{P}^{T/2}_{zz,k|k-1}$, can be evaluated. The estimated state vector and the square-root factor of the error covariance matrix required for SRCIF prediction can be recovered by using (6.136), (6.22) and (6.23). The derived SRCIF can then be extended to the case of multiple sensor. The prediction stage of the SRCIF in the multi-sensor state estimation is similar to that of SRCIF with a single sensor.

By using (6.141), (6.137) and (6.100), the multi-sensor SRCIF measurement update step can be derived as

$$\begin{bmatrix} \mathbf{Y}_{S,k|k-1} & \mathbf{Y}_{1,M,k|k-1} & \mathbf{Y}_{2,M,k|k-1} & \cdots & \mathbf{Y}_{D,M,k|k-1} \\ \hat{\mathbf{y}}^T_{S,k|k-1} & \mathbf{z}^T_{1,k}\mathbf{Y}_{\mathbf{R},1,k} & \mathbf{z}^T_{2,k}\mathbf{Y}_{\mathbf{R},2,k} & \cdots & \mathbf{z}^T_{D,k}\mathbf{Y}_{\mathbf{R},D,k} \end{bmatrix} \Theta = \begin{bmatrix} \mathbf{Y}_{S,k|k} & \mathbf{0} \\ \hat{\mathbf{y}}^T_{S,k|k} & \star \end{bmatrix}. \tag{6.120}$$

Squaring both sides of the above equation yields

$$\begin{bmatrix} \mathbf{Y}_{S,k|k-1} & \mathbf{Y}_{1,M,k|k-1} & \mathbf{Y}_{2,M,k|k-1} & \cdots & \mathbf{Y}_{D,M,k|k-1} \\ \hat{\mathbf{y}}^T_{S,k|k-1} & \mathbf{z}^T_{1,k}\mathbf{Y}_{\mathbf{R},1,k} & \mathbf{z}^T_{2,k}\mathbf{Y}_{\mathbf{R},2,k} & \cdots & \mathbf{z}^T_{D,k}\mathbf{Y}_{\mathbf{R},D,k} \end{bmatrix} \Theta\Theta^T \begin{bmatrix} \mathbf{Y}^T_{S,k|k-1} & \hat{\mathbf{y}}_{S,k|k-1} \\ \mathbf{Y}^T_{1,M,k|k-1} & \mathbf{Y}^T_{\mathbf{R},1,k}\mathbf{z}_{1,k} \\ \mathbf{Y}^T_{2,M,k|k-1} & \mathbf{Y}^T_{\mathbf{R},2,k}\mathbf{z}_{2,k} \\ \vdots & \vdots \\ \mathbf{Y}^T_{D,M,k|k-1} & \mathbf{Y}^T_{\mathbf{R},1,k}\mathbf{z}_{D,k} \end{bmatrix}$$

[4]**QR** is the orthogonal triangular decomposition and can be found in MATLAB using the command 'qr'.

$$= \begin{bmatrix} \mathbf{Y}_{S,k|k} & \mathbf{0} \\ \hat{\mathbf{y}}_{S,k|k}^T & \star \end{bmatrix} \begin{bmatrix} \mathbf{Y}_{S,k|k}^T & \hat{\mathbf{y}}_{S,k|k} \\ \mathbf{0} & \star \end{bmatrix} \quad (6.121)$$

$$\begin{bmatrix} \sum_{j=1}^{D} \mathbf{Y}_{j,M,k|k-1}\mathbf{Y}_{j,M,k|k-1}^T + \mathbf{Y}_{S,k|k-1}\mathbf{Y}_{S,k|k-1}^T & \sum_{j=1}^{D} \mathbf{Y}_{j,M,k|k-1}\mathbf{Y}_{R,j,k}^T \mathbf{z}_{j,k} + \mathbf{Y}_{S,k|k-1}\hat{\mathbf{y}}_{S,k|k-1} \\ \sum_{j=1}^{D} \mathbf{z}_{j,k}^T \mathbf{Y}_{R,j,k}\mathbf{Y}_{j,M,k|k-1}^T + \hat{\mathbf{y}}_{S,k|k-1}^T \mathbf{Y}_{S,k|k-1}^T & \sum_{j=1}^{D} \mathbf{z}_{j,k}\mathbf{Y}_{R,j,k}\mathbf{Y}_{R,j,k}^T \mathbf{z}_{j,k} + \hat{\mathbf{y}}_{S,k|k-1}^T \mathbf{Y}_{S,k|k-1}^T \end{bmatrix}$$

$$= \begin{bmatrix} \mathbf{Y}_{S,k|k}\mathbf{Y}_{S,k|k}^T & \mathbf{Y}_{S,k|k}\hat{\mathbf{y}}_{S,k|k} \\ \hat{\mathbf{y}}_{S,k|k}^T \mathbf{Y}_{S,k|k}^T & \star \end{bmatrix}. \quad (6.122)$$

Algorithm 6.5 Square-root cubature information filter

Prediction

1: Evaluate the cubature points

$$\chi_{i,k-1|k-1} = \mathbf{P}_{k-1|k-1}^{1/2}\xi_i + \mathbf{x}_{k-1|k-1}. \quad (6.123)$$

2: Evaluate the propagated cubature points

$$\chi_{i,k|k-1}^* = \mathbf{f}(\chi_{i,k-1|k-1}, \mathbf{u}_{k-1}). \quad (6.124)$$

3: Estimate the predicted state

$$\hat{\mathbf{x}}_{k|k-1} = \frac{1}{2n}\sum_{i=1}^{2n} \chi_{i,k|k-1}^*. \quad (6.125)$$

4: Estimate the square-root factor of the predicted error covariance and information matrix

$$\mathbf{S}_{k|k-1} = \left[\mathbf{qr}\left(\mathcal{X}_{i,k|k-1}^* \quad \mathbf{Y}_{Q,k-1}\right)^T\right]^T \quad (6.126)$$

$$\mathbf{Y}_{S,k|k-1} = \mathbf{S}_{k|k-1} \backslash \mathbf{I} \quad (6.127)$$

where $\mathbf{Y}_{Q,k-1}$ is a square-root factor of $\mathbf{Q}_{k-1}$ and

$$\mathcal{X}_{i,k|k-1}^* = \frac{1}{\sqrt{2n}}\left[\chi_{1,k|k-1}^* - \hat{\mathbf{x}}_{k|k-1} \quad \chi_{2,k|k-1}^* - \hat{\mathbf{x}}_{k|k-1} \quad \chi_{2n,k|k-1}^* - \hat{\mathbf{x}}_{k|k-1}\right]. \quad (6.128)$$

5: Evaluate the square-root information state vector

$$\hat{\mathbf{y}}_{S,k|k-1} = \mathbf{Y}_{S,k|k-1}\hat{\mathbf{x}}_{k|k-1}. \quad (6.129)$$

Measurement update

From (6.96), one can see that the measurement update stage of SREIF requires the linearised measurement model, $\nabla\mathbf{H}$. By using the statistical error propagation property below Sibley et al. (2006), the derivative-less measurement update step for the SRCIF can be derived.

$$\begin{aligned}\mathbf{P}_{zz} &= cov(\mathbf{z}) \\ &= E[(\mathbf{z}-\hat{\mathbf{z}})(\mathbf{z}-\hat{\mathbf{z}})^T] \\ &= E[\nabla\mathbf{h}_x\mathbf{x}-\nabla\mathbf{h}_x\hat{\mathbf{x}})(\nabla\mathbf{h}_x\mathbf{x}-\nabla\mathbf{h}_x\hat{\mathbf{x}})^T] \\ &= \nabla\mathbf{h}_x cov(\mathbf{x})\nabla\mathbf{h}_x^T = \nabla\mathbf{h}_x\mathbf{P}\nabla\mathbf{h}_x^T.\end{aligned} \tag{6.130}$$

Let $\mathbf{P}_{zz}^{1/2}$ and $\mathbf{P}^{1/2}$ denote the square root factor of $\mathbf{P}_{zz}$ and $\mathbf{P}$. Then (6.137) can be expressed as

$$\mathbf{P}_{zz}^{1/2}\mathbf{P}_{zz}^{T/2} = \nabla\mathbf{h}_x\mathbf{P}^{1/2}\mathbf{P}^{T/2}\nabla\mathbf{h}_x^T \tag{6.131}$$

and the covariance matrix in (6.131) can be factorised as

$$\mathbf{P}_{zz}^{1/2} = \nabla\mathbf{h}_x\mathbf{P}^{1/2} \tag{6.132}$$

$$\mathbf{P}_{zz}^{T/2} = \mathbf{P}^{T/2}\nabla\mathbf{h}_x^T. \tag{6.133}$$

Pre-multiplying $\nabla\mathbf{h}_x^T\mathbf{Y}_R$ by $\mathbf{P}_{k|k-1}^{-T/2}\mathbf{P}_{k|k-1}^{T/2}$ gives

$$\begin{aligned}\nabla\mathbf{h}_x^T\mathbf{Y}_R &= \mathbf{P}_{k|k-1}^{-T/2}\mathbf{P}_{k|k-1}^{T/2}\nabla\mathbf{h}_x^T\mathbf{Y}_R \\ &= \mathbf{P}_{k|k-1}^{-T/2}\mathbf{P}_{zz,k|k-1}^{T/2}\mathbf{Y}_R.\end{aligned} \tag{6.134}$$

From (6.97) and (6.141) it can be derived that

$$\begin{bmatrix}\mathbf{Y}_{S,k|k-1} & \mathbf{Y}_{M,k|k-1} \\ \hat{\mathbf{y}}_{S,k|k-1}^T & \mathbf{z}_k^T\mathbf{Y}_R\end{bmatrix}\Theta = \begin{bmatrix}\mathbf{Y}_{S,k|k} & \mathbf{0} \\ \hat{\mathbf{y}}_{S,k|k}^T & \star\end{bmatrix} \tag{6.135}$$

where

$$\mathbf{Y}_{M,k|k-1} = \mathbf{P}_{k|k-1}^{-T/2}\mathbf{P}_{zz,k|k-1}^{T/2}\mathbf{Y}_R. \tag{6.136}$$

6.4.6 Square-Root Cubature $\mathcal{H}_\infty$ Filter

In this subsection, the derivative-free, square-root cubature $\mathcal{H}_\infty$ filter for nonlinear systems is to be derived. The square-root cubature Kalman filter is a nonlinear filter which deals with only Gaussian noises (Arasaratnam and Haykin 2009) and the square-root $\mathcal{H}_\infty$ filter detailed in Sect. 6.4.3 can deal with non-Gaussian noises but it requires Jacobians during the nonlinear state estimation. The algorithm described here is derived from the square-root cubature Kalman filter and square-root $\mathcal{H}_\infty$ filter, and is called as the square-root cubature $\mathcal{H}_\infty$ filter (SRC$\mathcal{H}_\infty$F). The main advantages of the square-root cubature $\mathcal{H}_\infty$ filter are the avoidance of using derivatives in the nonlinear estimation and its capability to deal with non-Gaussian noises, in addition to the ability to alleviate influences of possible perturbations in system dynamics.

The prediction step of the square-root cubature $\mathcal{H}_\infty$ filter is the same as that of the square-root CKF (Arasaratnam and Haykin 2009).

SRC$\mathcal{H}_\infty$F Measurement Update

From (6.103), one can see that the measurement update stage of the square-root extended $\mathcal{H}_\infty$ filter requires the linearised measurement model, $\nabla\mathbf{h}$. By using the statistical error propagation property below (Sibley et al. 2006), the derivative-free measurement update step for the square-root cubature $\mathcal{H}_\infty$ filter can be derived as

$$\begin{aligned}\mathbf{P}_{zz} &\simeq \mathbb{E}[(\mathbf{z}-\hat{\mathbf{z}})(\mathbf{z}-\hat{\mathbf{z}})^T]\\ &\simeq \mathbb{E}[\nabla\mathbf{h}_x\mathbf{x}-\nabla\mathbf{h}_x\widehat{\mathbf{x}})(\nabla\mathbf{h}_x\mathbf{x}-\nabla\mathbf{h}_x\hat{\mathbf{x}})^T]\\ &\simeq \nabla\mathbf{h}_x cov(\mathbf{x})\nabla\mathbf{h}_x^T\\ &\simeq \nabla\mathbf{h}_x\mathbf{P}\nabla\mathbf{h}_x^T\end{aligned} \tag{6.137}$$

Let $\mathbf{P}_{zz}^{1/2}$ and $\mathbf{P}^{1/2}$ be the square root factors of $\mathbf{P}_{zz}$ and $\mathbf{P}$, respectively. Then (6.137) can be expressed as

$$\mathbf{P}_{zz}^{1/2}\mathbf{P}_{zz}^{T/2} = \nabla\mathbf{h}_x\mathbf{P}^{1/2}\mathbf{P}^{T/2}\nabla\mathbf{h}_x^T \tag{6.138}$$

and further (6.138) can be written as

$$\mathbf{P}_{zz}^{1/2} = \nabla\mathbf{h}_x\mathbf{P}^{1/2} \tag{6.139}$$

$$\mathbf{P}_{zz}^{T/2} = \mathbf{P}^{T/2}\nabla\mathbf{h}_x^T \tag{6.140}$$

By pre-multiplying $\mathbf{P}_{k|k-1}^{-T/2}\mathbf{P}_{k|k-1}^{T/2}$ to $\nabla\mathbf{h}_x^T\mathbf{R}_k^{-T/2}$ and by using (6.140), it can be obtained that

$$\begin{aligned}\nabla\mathbf{h}_x^T\mathbf{R}_k^{-T/2} &= \mathbf{P}_{k|k-1}^{-T/2}\mathbf{P}_{k|k-1}^{T/2}\nabla\mathbf{h}_x^T\mathbf{R}_k^{-T/2}\\ &= \mathbf{P}_{k|k-1}^{-T/2}\mathbf{P}_{zz,k|k-1}^{T/2}\mathbf{R}_k^{-T/2}\end{aligned} \tag{6.141}$$

By using (6.139) and (6.141) in (6.103), the measurement update of the square-root cubature $\mathcal{H}_\infty$ filter can be written as

$$\begin{bmatrix} -\mathbf{P}_{k|k-1}^{-T/2}\mathbf{P}_{zz,k|k-1}^{T/2}\mathbf{R}_k^{-T/2} & \mathbf{P}_{k|k-1}^{-T/2} & \gamma^{-1}\mathbf{I}_n \\ \mathbf{R}_k^{1/2} & \mathbf{P}_{zz,k|k-1}^{1/2} & \mathbf{0} \end{bmatrix}\Theta_J = \begin{bmatrix} \mathbf{0} & \mathbf{P}_{k|k}^{-T/2} & \mathbf{0} \\ \mathbf{R}_e^{1/2} & \mathbf{0} & \mathbf{0} \end{bmatrix} \tag{6.142}$$

The algorithm for the square-root cubature $\mathcal{H}_\infty$ filter is summarised in Algorithm 6.6. The derivation of (6.142) is given below as well.

Derivation of Update Step in SRC$\mathcal{H}_\infty$F

By expanding the left-hand side of (6.142), it can be seen that

$$\begin{bmatrix} -\mathbf{P}_{k|k-1}^{-T/2}\mathbf{P}_{zz,k|k-1}^{T/2}\mathbf{R}^{-T/2} & \mathbf{P}_{k|k-1}^{-T/2} & \gamma^{-1}\mathbf{I}_n \\ \mathbf{R}^{1/2} & \mathbf{P}_{zz,k|k-1}^{1/2} & \mathbf{0} \end{bmatrix} \Theta_J \mathbf{J} \Theta_J^T \begin{bmatrix} -\mathbf{R}^{-1/2}\mathbf{P}_{zz,k|k-1}^{1/2}\mathbf{P}_{k|k-1}^{-1/2} & \mathbf{R}^{T/2} \\ \mathbf{P}_{k|k-1}^{-1/2} & \mathbf{P}_{zz,k|k-1}^{T/2} \\ \gamma^{-1}\mathbf{I}_n & 0 \end{bmatrix}$$
$$= \begin{bmatrix} \mathbf{0} & \mathbf{P}_{k|k}^{-T/2} & \mathbf{0} \\ \mathbf{R}_e^{1/2} & \mathbf{0} & \mathbf{0} \end{bmatrix} \mathbf{J} \begin{bmatrix} \mathbf{0} & \mathbf{R}_e^{T/2} \\ \mathbf{P}_{k|k}^{-1/2} & \mathbf{0} \\ \mathbf{0} & \mathbf{0} \end{bmatrix} \tag{6.143}$$

By using $\Theta_J \mathbf{J} \Theta_J^T = \mathbf{J}$ and (6.107); (6.143) can be written as

$$\begin{bmatrix} \mathbf{P}_{k|k-1}^{-1} + \mathbf{P}_{k|k-1}^{-T/2}\mathbf{P}_{zz,k|k-1}^{T/2}\mathbf{R}_k^{-1}\mathbf{P}_{zz,k|k-1}^{1/2}\mathbf{P}_{k|k-1}^{-1/2} - \gamma^{-2}\mathbf{I}_n & \mathbf{0} \\ \mathbf{0} & \mathbf{P}_{zz,k|k-1} + \mathbf{R}_k \end{bmatrix} = \begin{bmatrix} \mathbf{P}_{k|k}^{-1} & \mathbf{0} \\ \mathbf{0} & \mathbf{R}_e \end{bmatrix} \tag{6.144}$$

By using the statistical approximations (6.68) and (6.140), the following can be derived:

$$\begin{aligned} \mathbf{P}_{xz,k|k-1} &\simeq \mathbf{P}_{k|k-1}\nabla\mathbf{h}_x^T \\ &= \mathbf{P}_{k|k-1}^{1/2}\mathbf{P}_{k|k-1}^{T/2}\nabla\mathbf{h}_x^T \\ &= \mathbf{P}_{k|k-1}^{1/2}(\nabla\mathbf{h}_x\mathbf{P}_{k|k-1}^{1/2})^T \\ &= \mathbf{P}_{k|k-1}^{1/2}\mathbf{P}_{zz,k|k-1}^{T/2} \end{aligned} \tag{6.145}$$

By equating the corresponding terms of LHS and RHS of (6.144) and by using (6.145) in (6.144) it yields

$$\mathbf{P}_{k|k}^{-1} = \mathbf{P}_{k|k-1}^{-1} + \mathbf{P}_{k|k-1}^{-1}\mathbf{P}_{xz,k|k-1}\mathbf{R}_k^{-1}\mathbf{P}_{xz,k|k-1}^{T}\mathbf{P}_{k|k-1}^{-T} - \gamma^{-2}\mathbf{I}_n \tag{6.146}$$

and

$$\mathbf{R}_e = \mathbf{P}_{zz,k|k-1} + \mathbf{R}_k \tag{6.147}$$

which is same as (6.73) in the cubature $\mathcal{H}_\infty$ filter.

Algorithm 6.6 Square-root cubature $\mathcal{H}_\infty$ filter

---

Prediction

1: Initialise the state vector, $\mathbf{x}$, and the square-root of the auxiliary matrix, $\mathbf{P}^{1/2}$.
2: Evaluate the cubature points

$$\boldsymbol{X}_{i,k-1|k-1} = \mathbf{P}^{1/2}_{k-1|k-1}\xi_i + \mathbf{x}_{k-1|k-1} \tag{6.148}$$

3: Evaluate the propagated cubature points

$$\boldsymbol{X}^*_{i,k|k-1} = \mathbf{f}(\boldsymbol{X}_{i,k-1|k-1}, \mathbf{u}_{k-1}) \tag{6.149}$$

4: Estimate the predicted state

$$\widehat{\mathbf{x}}_{k|k-1} = \frac{1}{2n}\sum_{i=1}^{2n}\boldsymbol{X}^*_{i,k|k-1} \tag{6.150}$$

5: Estimate the square-root factor of the predicted auxiliary matrix using the $\mathbf{QR}$[5] decomposition

$$\mathbf{P}^{1/2}_{k|k-1} = \left[\mathbf{qr}\left(\mathcal{X}_{i,k|k-1} \qquad \mathbf{Q}^{1/2}\right)^T\right]^T \tag{6.151}$$

where

$$\mathcal{X}_{i,k|k-1} = \frac{1}{\sqrt{2n}}\left[\boldsymbol{X}_{1,k|k-1} - \widehat{\mathbf{x}}_{k|k-1} \;\; \boldsymbol{X}_{2,k|k-1} - \widehat{\mathbf{x}}_{k|k-1} \ldots \boldsymbol{X}_{2n,k|k-1} - \widehat{\mathbf{x}}_{k|k-1}\right] \tag{6.152}$$

Measurement update

1: Evaluate the cubature points

$$\boldsymbol{X}_{i,k|k-1} = \mathbf{P}^{1/2}_{k|k-1}\xi_i + \widehat{\mathbf{x}}_{k|k-1}. \tag{6.153}$$

2: Evaluate the propagated cubature points

$$\mathbf{Z}_{i,k|k-1} = \mathbf{h}(\boldsymbol{X}_{i,k|k-1}, \mathbf{u}_k) \tag{6.154}$$

3: Evaluate the square-root factor of measurement auxiliary matrix

$$\mathbf{P}^{1/2}_{zz,k|k-1} = \left[\mathbf{qr}\left(\mathcal{Z}_{i,k|k-1} \qquad \mathbf{R}^{1/2}_k\right)^T\right]^T \tag{6.155}$$

where

$$\mathcal{Z}_{i,k|k-1} = \frac{1}{\sqrt{2n}}\left[\mathbf{Z}_{1,k|k-1} - \widehat{\mathbf{z}}_{k|k-1} \;\; \mathbf{Z}_{2,k|k-1} - \widehat{\mathbf{z}}_{k|k-1} \;\; \boldsymbol{X}_{2n,k|k-1} - \widehat{\mathbf{z}}_{k|k-1}\right] \tag{6.156}$$

4: Evaluate the square root of updated auxiliary matrix using

$$\begin{bmatrix} -\mathbf{P}_{k|k-1}^{-T/2}\mathbf{P}_{zz,k|k-1}^{T/2}\mathbf{R}_k^{-T/2} & \mathbf{P}_{k|k-1}^{-T/2} & \gamma^{-1}\mathbf{I}_n \\ \mathbf{R}_k^{1/2} & \mathbf{P}_{zz,k|k-1}^{1/2} & \mathbf{0} \end{bmatrix} \Theta_J = \begin{bmatrix} \mathbf{0} & \mathbf{P}_{k|k}^{-T/2} & \mathbf{0} \\ \mathbf{R}_e^{1/2} & \mathbf{0} & \mathbf{0} \end{bmatrix} \tag{6.157}$$

5: Evaluate the gain matrix using

$$\mathbf{K}_{SC\infty,k} = \mathbf{P}_{xz,k|k-1}\mathbf{R}_e^{-T/2}\mathbf{R}_e^{-1/2} \tag{6.158}$$

where

$$\mathbf{P}_{xz,k|k-1} = \mathbf{P}_{k|k-1}^{1/2}\mathbf{P}_{zz,k|k-1}^{T/2}$$

6: Evaluate the updated state using

$$\hat{\mathbf{x}}_{k|k} = \hat{\mathbf{x}}_{k|k-1} + \mathbf{K}_{SC\infty,k}(\mathbf{z}_k - \hat{\mathbf{z}}_{k|k-1}) \tag{6.159}$$

7: Recover the square-root factor of the updated auxiliary matrix, $\mathbf{P}_{k|k}^{1/2}$, using

$$\mathbf{P}_{k|k}^{1/2} = \mathbf{P}_{k|k}^{-1/2}\backslash\mathbf{I}_n. \tag{6.160}$$

6.4.7 Square-Root Cubature $\mathcal{H}_\infty$ Information Filter

In this subsection, a numerically stable square-root version of the cubature $\mathcal{H}_\infty$ information filter (SRC$\mathcal{H}_\infty$IF) is to be developed. Square-root filters arise when the covariance or information matrices are replaced by their square-root factors; these square root factors are then propagated at various stages. Square-root factors for the information matrix and other matrices are defined such that

$$\mathcal{I} = \mathcal{I}_s\mathcal{I}_s^T \tag{6.161}$$

$$\mathbf{P}_{xz} = \mathbf{P}_{xz,s}\mathbf{P}_{xz,s}^T \tag{6.162}$$

$$\mathbf{R}^{-1} = \mathbf{R}_{si}\mathbf{R}_{si}^T \tag{6.163}$$

$$\mathbf{Q} = \mathbf{Q}_s\mathbf{Q}_s^T \tag{6.164}$$

where $\mathcal{I}_s = \mathbf{P}^{\frac{-T}{2}}$, $\mathbf{P}_{xz,s} = \mathbf{P}_{xz,s}^{\frac{1}{2}}$, $\mathbf{R}_{si} = R^{\frac{-T}{2}}$, and $\mathbf{Q}_s = \mathbf{Q}^{\frac{1}{2}}$. Note that these square-root factors are not unique and can be calculated using different numerical techniques such as the QR decomposition, etc.

The square-root cubature $\mathcal{H}_\infty$ information filter (SRC$\mathcal{H}_\infty$IF) also comprises prediction and update stages. The prediction stage in SRC$\mathcal{H}_\infty$IF is the same as that of the square-root CKF, see Sect. 6.4.4 earlier in this chapter, also in Arasaratnam and Haykin (2009), but the information vector and the square root of the information matrix are propagated rather than the state and covariance matrix in the updating procedure.

The prediction stage of SRC$\mathcal{H}_\infty$IF has straight forward equations, please see the prediction stage of Algorithm 6.6 for more details. However, the update stage is not that straight forward. The update stage for the square-root cubature $\mathcal{H}_\infty$ information filter using the J-orthogonal transformation is given in the theorem below.

Theorem 6.1 *The updated information vector and information matrix for the SRC$\mathcal{H}_\infty$IF can be obtained from*

$$\begin{bmatrix} \mathcal{I}_{s,k|k-1}\boldsymbol{P}_{zz,s,k|k-1}\boldsymbol{R}_{si,k} & \mathcal{I}_{s,k|k-1} & \gamma^{-1}\mathbb{I}_n \\ \mathbf{z}_{a,k}^T\boldsymbol{R}_{si,k} & \mathfrak{s}_{s,k|k-1}^T & \boldsymbol{0} \end{bmatrix} \Theta_J = \begin{bmatrix} \mathcal{I}_{s,k|k} & \boldsymbol{0} & \boldsymbol{0} \\ \mathfrak{s}_{s,k|k}^T & \boldsymbol{0} & \star \end{bmatrix}, \tag{6.165}$$

where $\star$ represent the terms which are irrelevant for the SRC$\mathcal{H}_\infty$IF, $\mathbf{z}_{a,k} = \nu_k + \boldsymbol{P}_{zz,s,k|k-1}^T\mathcal{I}_{s,k|k-1}^T\boldsymbol{x}_{k|k-1}$ and Θ_J is a J-unitary matrix which satisfies

$$\Theta_J\Theta_J^T = J \quad and \quad J = \begin{bmatrix} \mathbb{I}_n & \boldsymbol{0} & \boldsymbol{0} \\ \boldsymbol{0} & \mathbb{I}_n & \boldsymbol{0} \\ \boldsymbol{0} & \boldsymbol{0} & -\mathbb{I}_n \end{bmatrix}.$$

Proof Squaring the left-hand side (LHS) of (6.165) yields

$$\begin{bmatrix} \mathcal{I}_{s,k|k-1}\mathbf{P}_{zz,s,k|k-1}\mathbf{R}_{si,k} & \mathcal{I}_{s,k|k-1} & \gamma^{-1}\mathbb{I}_n \\ \mathbf{z}_{a,k}^T\mathbf{R}_{si,k} & \mathfrak{s}_{s,k|k-1}^T & \mathbf{0} \end{bmatrix} \Theta_J\Theta_J^T \begin{bmatrix} \mathbf{R}_{si,k}^T\mathbf{P}_{zz,s,k|k-1}^T\mathcal{I}_{s,k|k-1}^T & \mathbf{R}_{si,k}^T\mathbf{z}_{a,k} \\ \mathcal{I}_{s,k|k-1}^T & \mathfrak{s}_{s,k|k-1} \\ \gamma^{-1}\mathbb{I}_n & \mathbf{0} \end{bmatrix}, \tag{6.166}$$

Further, (6.166) can be written as

$$\begin{aligned} &\begin{bmatrix} \mathcal{I}_{s,k|k-1}\mathbf{P}_{zz,s,k|k-1}\mathbf{R}_{si,k} & \mathcal{I}_{s,k|k-1} & \gamma^{-1}\mathbb{I}_n \\ \mathbf{z}_{a,k}^T\mathbf{R}_{si,k} & \mathfrak{s}_{s,k|k-1}^T & \mathbf{0} \end{bmatrix} \begin{bmatrix} \mathbb{I}_n & \mathbf{0} & \mathbf{0} \\ \mathbf{0} & \mathbb{I}_n & \mathbf{0} \\ \mathbf{0} & \mathbf{0} & -\mathbb{I}_n \end{bmatrix} \begin{bmatrix} \mathbf{R}_{si,k}^T\mathbf{P}_{zz,s,k|k-1}^T\mathcal{I}_{s,k|k-1}^T & \mathbf{R}_{si,k}^T\mathbf{z}_{a,k} \\ \mathcal{I}_{s,k|k-1}^T & \mathfrak{s}_{s,k|k-1} \\ \gamma^{-1}\mathbb{I}_n & \mathbf{0} \end{bmatrix}, \\ &= \begin{bmatrix} \left(\mathcal{I}_{s,k|k-1}\mathbf{P}_{zz,s,k|k-1}\mathbf{R}_k^{-1}\mathbf{P}_{zz,s,k|k-1}^T\mathcal{I}_{s,k|k-1}^T + \mathcal{I}_{k|k-1} - \gamma^{-2}\mathbb{I}_n\right) & \left(\mathcal{I}_{s,k|k-1}\mathbf{P}_{zz,s,k|k-1}\mathbf{R}_k^{-1}\mathbf{z}_{a,k} + \mathcal{I}_{s,k|k-1}\mathfrak{s}_{s,k|k-1}\right) \\ \star & \star \end{bmatrix} \end{aligned} \tag{6.167}$$

The covariance matrix in (6.81) can be factorised as

$$\begin{aligned} \mathbf{P}_{zz,k|k-1} &= \nabla\mathbf{h}\mathbf{P}_{k|k-1}\nabla\mathbf{h}^T \\ &= \nabla\mathbf{h}\mathbf{P}_{k|k-1}^{\frac{1}{2}}\mathbf{P}_{k|k-1}^{\frac{T}{2}}\nabla\mathbf{h}^T \\ &= (\nabla\mathbf{h}\mathbf{P}_{s,k|k-1})(\nabla\mathbf{h}\mathbf{P}_{s,k|k-1})^T \\ &= \mathbf{P}_{zz,s,k|k-1}\mathbf{P}_{zz,s,k|k-1}^T. \end{aligned} \tag{6.168}$$

The $\mathbf{P}_{xz,k|k-1}$ in (6.82) can be represented in terms of $\mathbf{P}_{zz,s,k|k-1}$ and $\mathbf{P}_{s,k|k-1}$ as

$$\begin{aligned}\mathbf{P}_{xz,k|k-1} &= \mathbf{P}_{k|k-1}\nabla\mathbf{h}^T \\ &= \mathbf{P}_{k|k-1}^{\frac{1}{2}}(\nabla\mathbf{h}\mathbf{P}_{k|k-1}^{\frac{1}{2}})^T \\ &= \mathbf{P}_{s,k|k-1}\mathbf{P}_{zz,s,k|k-1}^T \end{aligned} \tag{6.169}$$

Pre-multiplying the information matrix on both sides of the aforementioned equation yields

$$\begin{aligned}\mathcal{I}_{k|k-1}\mathbf{P}_{xz,k|k-1} &= \mathcal{I}_{k|k-1}\mathbf{P}_{s,k|k-1}\mathbf{P}_{zz,s,k|k-1}^T \\ &= \mathcal{I}_{s,k|k-1}\mathcal{I}_{s,k|k-1}^T\mathbf{P}_{s,k|k-1}\mathbf{P}_{zz,s,k|k-1}^T \\ &= \mathcal{I}_{s,k|k-1}\mathbf{P}_{zz,s,k|k-1}^T \end{aligned} \tag{6.170}$$

Substituting (6.170) in (6.167) yields

$$\begin{aligned}&\begin{bmatrix} \left(\mathcal{I}_{k|k-1}\mathbf{P}_{xz,k|k-1}\mathbf{R}_k^{-1}\mathbf{P}_{xz,k|k-1}^T\mathcal{I}_{k|k-1}^T + \mathcal{I}_{k|k-1} - \gamma^{-2}\mathbb{I}_n\right) & \left(\mathcal{I}_{k|k-1}\mathbf{P}_{xz,k|k-1}\mathbf{R}_k^{-1}\mathbf{z}_{a,k} + \mathcal{I}_{s,k|k-1}\mathfrak{s}_{s,k|k-1}\right) \\ \star & \star \end{bmatrix} \\ &= \begin{bmatrix} \left(\mathcal{I}_{k|k-1} + \mathcal{I}_{k|k-1}\mathbf{P}_{xz,k|k-1}\mathbf{R}_k^{-1}\mathbf{P}_{xz,k|k-1}^T\mathcal{I}_{k|k-1}^T - \gamma^{-2}\mathbb{I}_n\right) & \left(\mathfrak{s}_{k|k-1} + \mathcal{I}_{k|k-1}\mathbf{P}_{xz,k|k-1}\mathbf{R}_k^{-1}\mathbf{z}_{a,k}\right) \\ \star & \star \end{bmatrix}. \end{aligned} \tag{6.171}$$

Now by squaring the right-hand side (RHS) of the update stage in (6.165), the following can be derived:

$$\begin{aligned}&\begin{bmatrix} \mathcal{I}_{s,k|k} & \mathbf{0} & \mathbf{0} \\ \mathfrak{s}_{s,k|k}^T & \mathbf{0} & \star \end{bmatrix}\begin{bmatrix} \mathcal{I}_{s,k|k}^T & \mathfrak{s}_{s,k|k} \\ \mathbf{0} & \mathbf{0} \\ \mathbf{0} & \star \end{bmatrix} \\ &= \begin{bmatrix} \mathcal{I}_{k|k}\mathcal{I}_{s,k|k}^T & \mathcal{I}_{s,k|k}\mathfrak{s}_{s,k|k} \\ \star & \star \end{bmatrix} \\ &= \begin{bmatrix} \mathcal{I}_{s,k|k} & \mathfrak{s}_{k|k} \\ \star & \star \end{bmatrix}. \end{aligned} \tag{6.172}$$

By equating the corresponding elements of (6.171) and (6.172), which corresponds to the LHS and RHS of (6.165), one gets

$$\mathfrak{s}_{k|k} = \mathfrak{s}_{k|k-1} + \mathbf{i}_k \tag{6.173}$$

$$\mathcal{I}_{k|k} = \mathcal{I}_{k|k-1} + \mathbf{I}_k. \tag{6.174}$$

where

$$\mathbf{i}_k = \mathcal{I}_{k|k-1}\mathbf{P}_{xz,k|k-1}\mathbf{R}_k^{-1}[\nu_k + \mathbf{P}_{xz,k|k-1}^T\mathcal{I}_{k|k-1}^T\mathbf{x}_{k|k-1}] \tag{6.175}$$

$$\mathbf{I}_k = \mathcal{I}_{k|k-1}\mathbf{P}_{xz,k|k-1}\mathbf{R}_k^{-1}\mathbf{P}_{xz,k|k-1}^T\mathcal{I}_{k|k-1}^T - \gamma^{-2}\mathbb{I}_n. \tag{6.176}$$

Note that the information vector and the information matrix given in (6.173) and (6.174), and the corresponding information contributions in (6.175) and (6.176) for the SRC$\mathcal{H}_\infty$IF are the same as those of the C$\mathcal{H}_\infty$IF in (6.79), (6.80), (6.84) and (6.85). This proves that the update stage of the SRC$\mathcal{H}_\infty$IF given in Theorem 6.1 is an equivalent of the C$\mathcal{H}_\infty$IF in Sect. 6.3.4.

The prediction stage of multi-sensor SRC$\mathcal{H}_\infty$IF will be the same as that of the single sensor SRC$\mathcal{H}_\infty$IF, since the multiple sensors only affect the update stage. The updated information vector and information matrix of the multi-sensor SRC$\mathcal{H}_\infty$IF can be obtained from

$$\begin{bmatrix} \mathcal{I}_{s,k|k-1}\mathbf{P}_{zz,1,s,k|k-1}\mathbf{R}_{si,1,k} & \mathcal{I}_{s,k|k-1}\mathbf{P}_{zz,2,s,k|k-1}\mathbf{R}_{si,2,k} & \dots & \mathcal{I}_{s,k|k-1}\mathbf{P}_{zz,D,s,k|k-1}\mathbf{R}_{si,D,k} & \mathcal{I}_{s,k|k-1} & \gamma^{-1}\mathbb{I} \\ \mathbf{z}_{a,1,k} & \mathbf{z}_{a,2,k} & \dots & \mathbf{z}_{a,D,k} & \mathfrak{s}_{k|k-1} & \mathbf{0} \end{bmatrix}\Theta_J$$
$$= \begin{bmatrix} \mathcal{I}_{s,k|k} & \mathbf{0} & \mathbf{0} \\ \mathfrak{s}_{s,k|k} & \mathbf{0} & \star \end{bmatrix} \tag{6.177}$$

The update stage for multi-sensor SRC$\mathcal{H}_\infty$IF in (6.177) can easily be proved along the similar lines of the proof of Theorem 6.1. The square-root cubature $\mathcal{H}_\infty$ information filters for the single sensor and multi-sensors are summarised in Algorithm 6.7. Apart from its applicability for the multi-sensor state estimations in the presence of Gaussian or non-Gaussian noises, SRC$\mathcal{H}_\infty$IF can also handle the case of ill-conditioned covariance matrices and it possesses double-order precision in the computation. As such, the square-root cubature $\mathcal{H}_\infty$ information filters are suitable for efficient, real-time implementations. They can also easily estimate the states with near-perfect measurements and/or in cases of systems with uncertainties up to a certain degree.

Algorithm 6.7 Square-root cubature $\mathcal{H}_\infty$ information filter

Initialise the information vector and square-root information matrix, $\mathcal{I}_{s,0|0}$ and $\mathfrak{s}_{s,0|0}$ for $k = 1$.

Prediction

1: Evaluate $\chi_{i,k-1|k-1}$, $\chi^*_{i,k|k-1}$ and $\mathbf{x}_{k|k-1}$ as

$$\begin{aligned} \chi_{i,k-1|k-1} &= \left(\mathcal{I}_{s,k-1|k-1}\backslash\mathbb{I}_n\right)\xi_i + \left(\mathcal{I}_{s,k-1|k-1}\backslash\mathfrak{s}_{s,k-1|k-1}\right), \\ \chi^*_{i,k|k-1} &= \mathbf{f}(\chi_{i,k-1|k-1}, \mathbf{u}_{k-1}) \\ \mathbf{x}_{k|k-1} &= \frac{1}{2n}\sum_{i=1}^{2n}\chi^*_{i,k|k-1} \end{aligned}$$

where '\' is the left-divide operator, $\mathbb{I}_n$ is the nth order identity matrix, and ξ_i is the i-th element of the following set:

$$\sqrt{n}\left[\begin{pmatrix}1\\0\\\vdots\\0\end{pmatrix}, \dots, \begin{pmatrix}0\\\vdots\\0\\1\end{pmatrix}, \begin{pmatrix}-1\\0\\\vdots\\0\end{pmatrix}, \dots, \begin{pmatrix}0\\\vdots\\0\\-1\end{pmatrix}\right] \tag{6.178}$$

2: Evaluate the information matrix

$$\mathcal{I}_{s,k|k-1} = \left[\mathbf{qr}\left(\mathbf{X}^{*}_{i,k|k-1} \quad \mathbf{Q}_{s,k-1}\right)^T\right]^T \Big\backslash \mathbb{I}_n$$

$$\mathfrak{s}_{s,k|k-1} = \mathcal{I}_{s,k|k-1}\mathbf{x}_{k|k-1}$$

where

$$\mathbf{X}^{*}_{i,k|k-1} = \frac{1}{\sqrt{2n}}\left[\chi^{*}_{1,k|k-1} - \hat{\mathbf{x}}_{k|k-1} \quad \chi^{*}_{2,k|k-1} - \hat{\mathbf{x}}_{k|k-1} \ldots \quad \chi^{*}_{2n,k|k-1} - \hat{\mathbf{x}}_{k|k-1}\right].$$

Measurement update

1: Evaluate $\chi_{i,k|k-1}$, $\mathbb{Z}_{i,k|k-1}$ and $\mathbf{z}_{k|k-1}$ as

$$\chi_{i,k|k-1} = \mathcal{I}_{s,k|k-1}\backslash\mathbb{I}_n\xi_i + \mathbf{x}_{k|k-1}$$

$$\mathbb{Z}_{i,k|k-1} = \mathbf{h}(\chi_{i,k|k-1}, \mathbf{u}_k)$$

$$\mathbf{z}_{k|k-1} = \frac{1}{2n}\sum_{i=1}^{2n}\mathbb{Z}_{i,k|k-1}.$$

2: Evaluate

$$\mathbf{P}_{zz,s,k|k-1} = \left[\mathbf{qr}\left(\mathbf{Z}_{i,k|k-1} \quad \mathbf{R}_{si,k}\right)^T\right]^T \Big\backslash$$

where

$$\mathbf{Z}_{i,k|k-1} = \frac{1}{\sqrt{2n}}\left[\mathbb{Z}_{1,k|k-1} - \mathbf{z}_{k|k-1} \quad \mathbb{Z}_{2,k|k-1} - \mathbf{z}_{k|k-1} \ldots \quad \mathbb{Z}_{2n,k|k-1} - \mathbf{z}_{k|k-1}\right].$$

3: For a single sensor, the square-root information matrix and the corresponding information vector can be obtained from

$$\begin{bmatrix} \mathcal{I}_{s,k|k-1}\mathbf{P}_{zz,s,k|k-1}\mathbf{R}_{si,k} & \mathcal{I}_{s,k|k-1} & \gamma^{-1}\mathbb{I}_n \\ \mathbf{z}^T_{a,k}\mathbf{R}_{si,k} & \mathfrak{s}^T_{s,k|k-1} & \mathbf{0} \end{bmatrix}\Theta_J = \begin{bmatrix} \mathcal{I}_{s,k|k} & \mathbf{0} & \mathbf{0} \\ \mathfrak{s}^T_{s,k|k} & \mathbf{0} & \star \end{bmatrix},$$

where γ is an attenuation parameter, Θ_J is a J-unitary matrix, $\mathbf{z}_{a,k} = \nu_k + \mathbf{P}^T_{zz,s,k|k-1}\mathcal{I}^T_{s,k|k-1}\mathbf{x}_{k|k-1}$, and $\star$ represent the terms which are irrelevant for the SR-C$\mathcal{H}_\infty$IF.

Measurement update for multi-sensor state estimation

1: For the case of multiple sensors, the square-root information matrix and the corresponding information vector can be obtained from

$$\begin{bmatrix} \mathcal{I}_{s,k|k-1}\mathbf{P}_{zz,1,s,k|k-1}\mathbf{R}_{si,1,k} & \mathcal{I}_{s,k|k-1}\mathbf{P}_{zz,2,s,k|k-1}\mathbf{R}_{si,2,k} & \cdots & \mathcal{I}_{s,k|k-1}\mathbf{P}_{zz,D,s,k|k-1}\mathbf{R}_{si,D,k} & \mathcal{I}_{s,k|k-1} & \gamma^{-1}\mathbb{I}_n \\ \mathbf{z}_{a,1,k} & \mathbf{z}_{a,2,k} & \cdots & \mathbf{z}_{a,D,k} & \mathfrak{s}_{k|k-1} & \mathbf{0} \end{bmatrix}\Theta_J = \begin{bmatrix} \mathcal{I}_{s,k|k} & \mathbf{0} & \mathbf{0} \\ \mathfrak{s}_{s,k|k} & \mathbf{0} & \star \end{bmatrix}$$

6.5 State Estimation of a Permanent Magnet Synchronous Motor

Before leaving this chapter, a near real-world case study will be investigated for application of various state estimation approached discussed earlier in the chapter. That plant under investigation is a two phase permanent synchronous motor system (PMSM) Simon (2006). The PMSM has four states, the first two states are currents through the two windings, the third state is the angular speed and the fourth state the rotor angular position. The inputs are two voltages. The objective is to estimate the rotor angular speed and position of PMSM with the two winding currents as measurements (the outputs). A number of state estimation approaches will be tested and compared, including the square-root cubature Kalman filter (SRCKF) and the square-root cubature $\mathcal{H}_\infty$ filter (SRC$\mathcal{H}_\infty$F), the square-root extended information filter (SREIF) and the square root cubature information filter (SRCIF) as well as the square-root cubature $\mathcal{H}_\infty$ information filter (SRC$\mathcal{H}_\infty$IF). Also, in the simulations, cases of multiple sensors are to be considered. Both Gaussian and non-Gaussian noises will be incorporated in the simulations.

The dynamics of the motor system will be introduced first, followed by applications of state estimation approaches.

6.5.1 PMSM Model

The discrete-time, nonlinear model of PMSM is as follows Simon (2006):

$$\begin{bmatrix} x_{1,k+1} \\ x_{2,k+1} \\ x_{3,k+1} \\ x_{4,k+1} \end{bmatrix} = \begin{bmatrix} x_{1,k} + T_s\left(-\frac{R}{L}x_{1,k} + \frac{\omega\lambda}{L}\sin x_{4,k} + \frac{1}{L}u_{a,k}\right) \\ x_{2,k} + T_s\left(-\frac{R}{L}x_{2,k} - \frac{\omega\lambda}{L}\cos x_{4,k} + \frac{1}{L}u_{b,k}\right) \\ x_{3,k} + T_s\left(-\frac{3\lambda}{2J}x_{1,k}\sin x_{4,k} + \frac{3\lambda}{2J}x_{2,k}\cos x_{4,k} - \frac{Fx_{3,k}}{J}\right) \\ x_{4,k} + T_s x_{3,k} \end{bmatrix}$$

The outputs and inputs are

$$\begin{bmatrix} y_{1,k} \\ y_{2,k} \end{bmatrix} = \begin{bmatrix} x_{1,k} \\ x_{2,k} \end{bmatrix}, \quad \begin{bmatrix} u_{1,k} \\ u_{2,k} \end{bmatrix} = \begin{bmatrix} \sin(0.002\pi k) \\ \cos(0.002\pi k) \end{bmatrix}.$$

The following parameters are considered in the estimator design and in the simulations: R $= 1.9\Omega$, $\lambda = 0.1$, L $= 0.003$ H, J $= 0.00018$, F $= 0.001$ and $T_s =$ 0.001 s.

6.5.2 State Estimation Using SREIF and SRCIF

The speed and the rotor angular position are now to be estimated using SREIF and SRCIF. The covariance matrices for the process and measurement noises

$$\mathbf{Q} = \begin{bmatrix} 11.11 & 0 & 0 & 0 \\ 0 & 11.11 & 0 & 0 \\ 0 & 0 & 0.25 & 0 \\ 0 & 0 & 0 & 1 \times 10^{-6} \end{bmatrix}, \quad \mathbf{R} = 1 \times 10^{-6}\mathbf{I}_2$$

are added to the plant and measurement models. The initial conditions for all the plant states are 0, the initial information vector is selected from $\mathcal{N}\left(\begin{bmatrix} 1 & 1 & 1 & 1 \end{bmatrix}^T, \mathbf{I}_4\right)$.

Over 500 Monte Carlo runs were performed to analyse the performance of the estimates. Figure 6.1 shows a typical result of one of the Monte Carlo simulations. In Fig. 6.1 the estimated states by SRCIF are very close to the actual states, whereas estimated states using SREIF fail to converge to the actual states. Note that the speed in Fig. 6.1 could be negative as the PMSM is simulated based on an open-loop control structure. The desired speed tracking can of course be achieved by designing a closed-loop control scheme. The root mean square error (RMSE) plot for the PMSM's speed is shown in Fig. 6.2. The peaks in Fig. 6.2 are due to the sudden change in the rotor angle which occurs at the rotor angle of 2π, where SREIF estimation performance is poor. The average RMSE values over the simulated time period are 8.1874 and 2.2909 for SREIF and SRCIF, respectively.

To show the effectiveness of the SRCIF in the case of multiple sensors, data from two different sets of sensors are used in the simulation. The noise covariances of the two sensors are

$$\mathbf{R}_1 = 1 \times 10^{-6} I_2 \quad \text{and} \quad \mathbf{R}_2 = 2.5 \times 10^{-5} I_2.$$

The simulation results for the state estimation of the PMSM using the multi-sensor SREIF and SRCIF are shown in Fig. 6.3. The first two actual states and their estimates using decentralised SREIF and SRCIF almost overlapped and hence they are not shown in Fig. 6.3. The corresponding RMSE plots are shown in Fig. 6.4.

Although the issues of fault detection and isolation are not within the main scope of this book, the state estimation in the presence of data loss is however simulated in this example. It is assumed that the first and second sensor fails to give any output from 0.5 to 0.7 s and from 2 to 2.5 s, respectively. It can be seen that the state estimates using multi-sensor SRCIF in the presence of data loss are very close to each other. The sudden changes in the rotor position and angle at 0.5 and 2 s are due to the data loss. The average RMSE values over the simulation time period are 8.0975 and 2.5747 for multi-sensor SREIF and SRCIF, respectively. In summary, the figures displayed simulation results from two scenarios: one with near-perfect measurements and the other with data loss in the measurements. It can be seen that, in the two different sensors, the multi-sensor SRCIF outperforms multi-sensor SREIF.

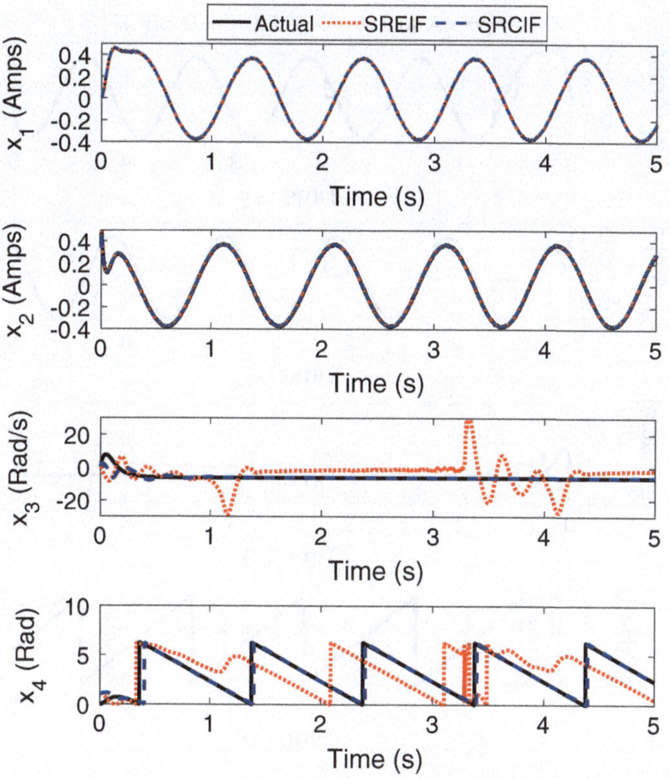

Fig. 6.1 Actual and estimated states of PMSM using SREIF and SRCIF

Fig. 6.2 RMSE of PMSM using SREIF and SRCIF

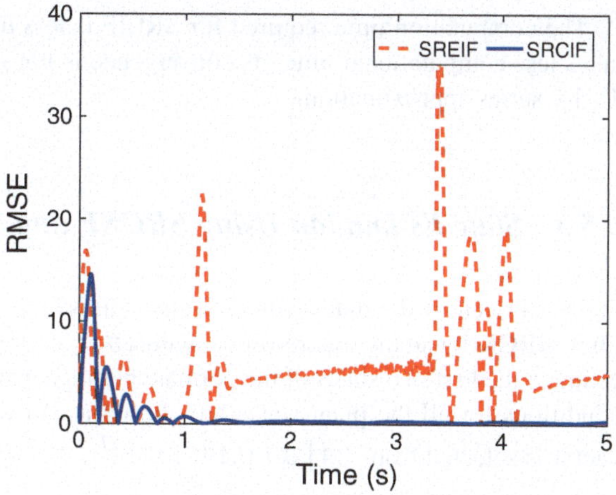

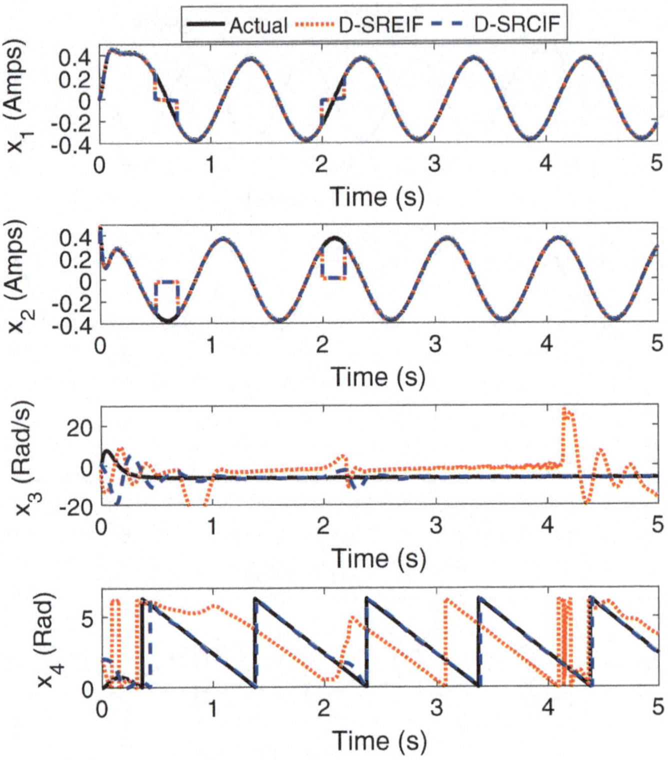

Fig. 6.3 Actual and estimated states using decentralised SREIF and SRCIF

The computation time required for SREIF is less than SRCIF. Although SREIF takes less computational time, its convergence is not guaranteed due the first-order Taylor series approximation.

6.5.3 State Estimation Using SRCKF and SRC$\mathcal{H}_\infty$F

In this subsection, the state estimation algorithms of the square root cubature Kalman filter (SRCKF) and the square-root cubature $\mathcal{H}_\infty$ filter (SRC$\mathcal{H}_\infty$F) will be applied to estimate the first two states of the permanent magnet synchronous motor. The initial conditions for all the plant states are chosen as 0.1 and the initial estimated state vector is selected from $\mathbb{U}\left(\left[0.1\ 0.1\ 0.1\ 0.1\right]^T, 0.1 \times \mathbf{I}_4\right)$.

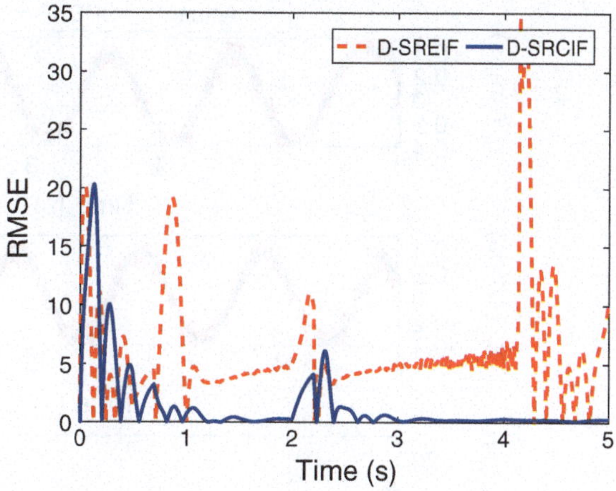

Fig. 6.4 RMSE of PMSM using decentralised SREIF and SRCIF

6.5.3.1 State Estimation in the Presence of Non-Gaussian Noise

In this case the process and measurement noises are assumed to be non-Gaussian uniform distributed. For the process noise, the standard deviations of the four states are 0.01, 0.01, 0.5 and 0.0001, and the standard deviations for the measurement noise are 0.05 and 0.05. The actual and estimated states, and the RMSEs using the square-root CKF and square root cubature $\mathcal{H}_\infty$ filter are shown in Figs. 6.5 and 6.6, respectively. The RMSE for the square-root CKF is higher than the square-root cubature $\mathcal{H}_\infty$ filter. The maximum RMSEs for the square-root CKF and the square-root cubature $\mathcal{H}_\infty$ filter are 0.7113 and 0.3679, respectively. Hence, for the PMSM also the square-root cubature $\mathcal{H}_\infty$ filter would appear to be well suited in the presence of non-Gaussian noises.

6.5.3.2 State Estimation with Near-Perfect Measurements

In this scenario, the simulations above are repeated with near-perfect measurements. Two groups of results are demonstrated: (a) with non-Gaussian process noise; and, (b) with Gaussian process noise. For the Gaussian case, the standard deviations of the four states for process noise are the same as that of the non-Gaussian case as given earlier. The standard deviations of the measurement noise are 1×10^{-15} and 1×10^{-15} for both Gaussian and non-Gaussian simulations. The corresponding results are shown in the Fig. 6.7. Similar to CSTR, the RMSEs using square-root cubature $\mathcal{H}_\infty$ filters in the presence of Gaussian and non-Gaussian are also within an acceptable range.

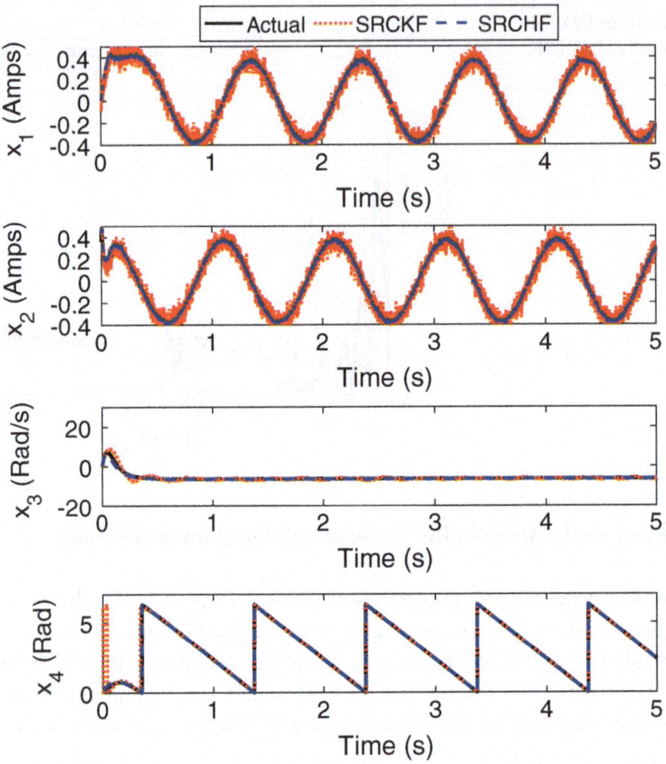

Fig. 6.5 Actual and estimated states of PMSM for non-Gaussian noise using SRCKF and SRC$\mathcal{H}_\infty$F

Fig. 6.6 RMSEs of the PMSM using SRCKF and SRC$\mathcal{H}_\infty$F

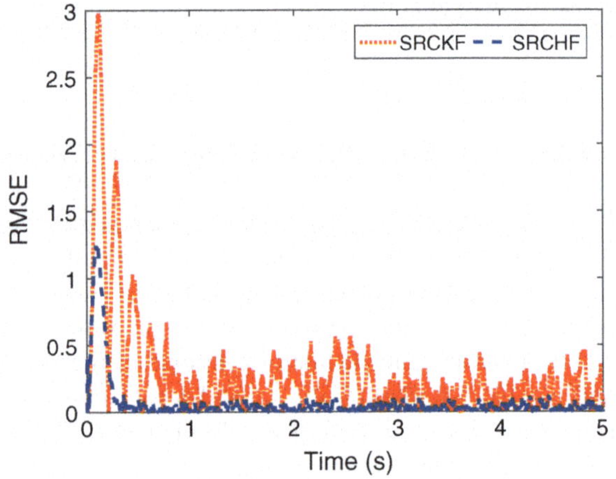

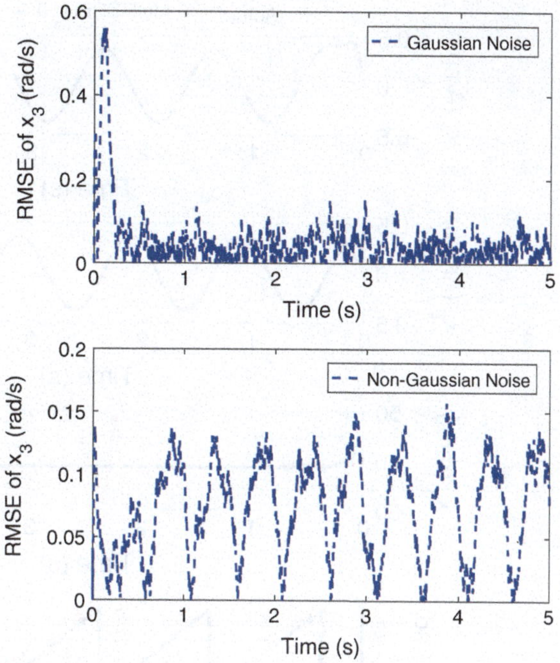

Fig. 6.7 RMSEs of x_3 with near-perfect measurements for the PMSM using SRCKF and SRC$\mathcal{H}_\infty$F

6.5.4 *State Estimation with Multi-sensors Using SREIF, SRCIF and SRC$\mathcal{H}_\infty$IF*

In this part, the states of PMSM are to be estimated by using multiple sensors. The approach applied is mainly the square-root cubature $\mathcal{H}_\infty$ information filter (SRC$\mathcal{H}_\infty$IF), though other filters will be tested in simulations for comparison purpose. Two scenarios are considered: (a) state estimation in the presence of Gaussian noise; and, (b) state estimation in the presence of non-Gaussian noise.

6.5.4.1 Multi-sensor State Estimation in the Presence of Low Gaussian Noise

It is first assumed that the process and measurement noises are Gaussian. For all the simulation results given in this part, the initial conditions for the actual PMSM states and the information vector are randomly selected from $\mathcal{N}([0.1 \quad 0.1 \quad 0.1 \quad 0.1]^T, 0.1 \times I_4)$.

SRC$\mathcal{H}_\infty$IF. For multi-sensor state estimation, two sets of current sensors are used to estimate the states. The process and measurement noise covariance matrices for are chosen as

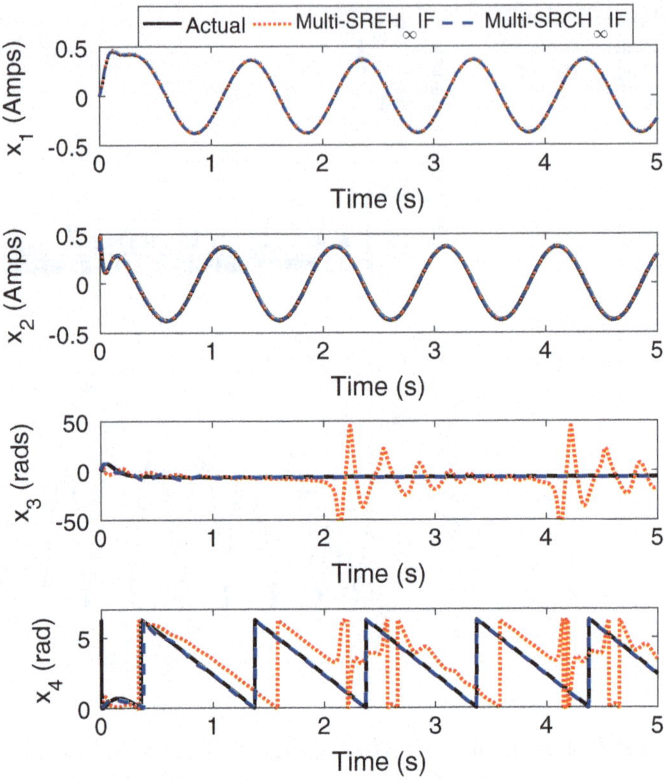

Fig. 6.8 PMSM states using decentralised SRE$\mathcal{H}_\infty$IF and decentralised SRC$\mathcal{H}_\infty$IF for low Gaussian noise

$$\mathbf{Q} = \begin{bmatrix} 6.25 & 0 & 0 & 0 \\ 0 & 6.25 & 0 & 0 \\ 0 & 0 & 0.1 & 0 \\ 0 & 0 & 0 & 1\times 10^{-6} \end{bmatrix}, \quad \mathbf{R}_1 = \begin{bmatrix} 2.5\times 10^{-6} & 0 \\ 0 & 2.5\times 10^{-6} \end{bmatrix} \quad \mathbf{R}_2 = \begin{bmatrix} 5\times 10^{-6} & 0 \\ 0 & 5\times 10^{-6} \end{bmatrix}$$

Simulation results of state estimation in the presence of low Gaussian noise using the multi-sensor SRE$\mathcal{H}_\infty$IF and SRC$\mathcal{H}_\infty$IF are shown in Fig. 6.8. The actual states and their estimates using the SRE$\mathcal{H}_\infty$IF and SRC$\mathcal{H}_\infty$IF are shown in Fig. 6.8. The estimates using the SRE$\mathcal{H}_\infty$IF are more erroneous than the SRC$\mathcal{H}_\infty$IF. The root means square errors (RMSEs) for the speed using SRE$\mathcal{H}_\infty$IF and SRC$\mathcal{H}_\infty$IF are shown in Fig. 6.9, where the SRE$\mathcal{H}_\infty$IF RMSE is larger than that of SRC$\mathcal{H}_\infty$IF. The maximum RMSE error is 58.6 for the SRE$\mathcal{H}_\infty$IF and 4.7 for the SRC$\mathcal{H}_\infty$IF, and the average RMSE error is 12.89 for the SRE$\mathcal{H}_\infty$IF and 1.21 for the SRC$\mathcal{H}_\infty$IF.

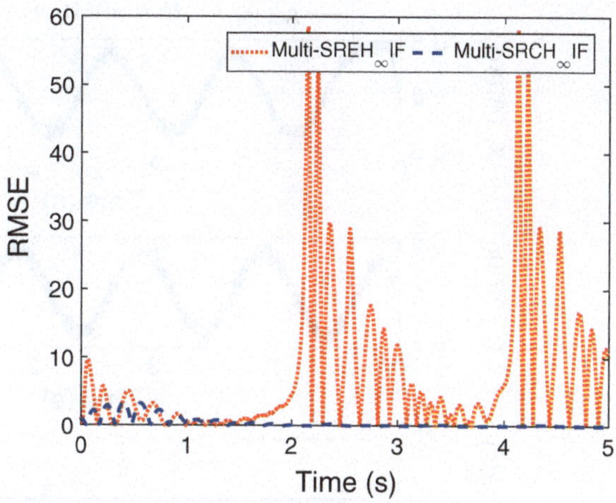

Fig. 6.9 RMSE of PMSM speed using decentralised SRE$\mathcal{H}_\infty$IF and decentralised SRC$\mathcal{H}_\infty$IF for low Gaussian noise

6.5.4.2 Multi-sensor State Estimation in the Presence of High Gaussian Noise

The simulations are now performed with high Gaussian noise. It is assumed that the sensors and plant model used here are more prone to noise such that their noise covariance matrices are assumed to be 12 times those given in last simulation with low Gaussian noise. The process and measurement noise covariance matrices for high Gaussian noise are chosen as

$$\mathbf{Q} = \begin{bmatrix} 75 & 0 & 0 & 0 \\ 0 & 75 & 0 & 0 \\ 0 & 0 & 1.2 & 0 \\ 0 & 0 & 0 & 1.2 \times 10^{-5} \end{bmatrix}, \quad \mathbf{R}_1 = \begin{bmatrix} 3 \times 10^{-5} & 0 \\ 0 & 3 \times 10^{-5} \end{bmatrix}, \quad \mathbf{R}_2 = \begin{bmatrix} 6 \times 10^{-5} & 0 \\ 0 & 6 \times 10^{-5} \end{bmatrix}.$$

Simulation results of a typical simulation run are shown in Fig. 6.10. The actual states and their estimates using the SRE$\mathcal{H}_\infty$IF and SRC$\mathcal{H}_\infty$IF are shown in Fig. 6.10. Due to the noisy current sensors, the first two states (the currents of PMSM) in Fig. 6.10 are noisy. The root means square errors (RMSEs) for the speed using SRE$\mathcal{H}_\infty$IF and SRC$\mathcal{H}_\infty$IF are shown in Fig. 6.11, where the SRE$\mathcal{H}_\infty$IF RMSE is larger than that of SRC$\mathcal{H}_\infty$IF. The maximum RMSE error is 58.95 for the SRE$\mathcal{H}_\infty$IF and 23.25 for SRC$\mathcal{H}_\infty$IF, and the average RMSE error is 13.59 for SRE$\mathcal{H}_\infty$IF and 4.09 for SRC$\mathcal{H}_\infty$IF.

6.5.4.3 Multi-sensor State Estimation in the Presence of Low Non-Gaussian Noises

In some of the control engineering applications, the process and/or measurement noises can be approximated by a Rayleigh probability distribution function Rigatos

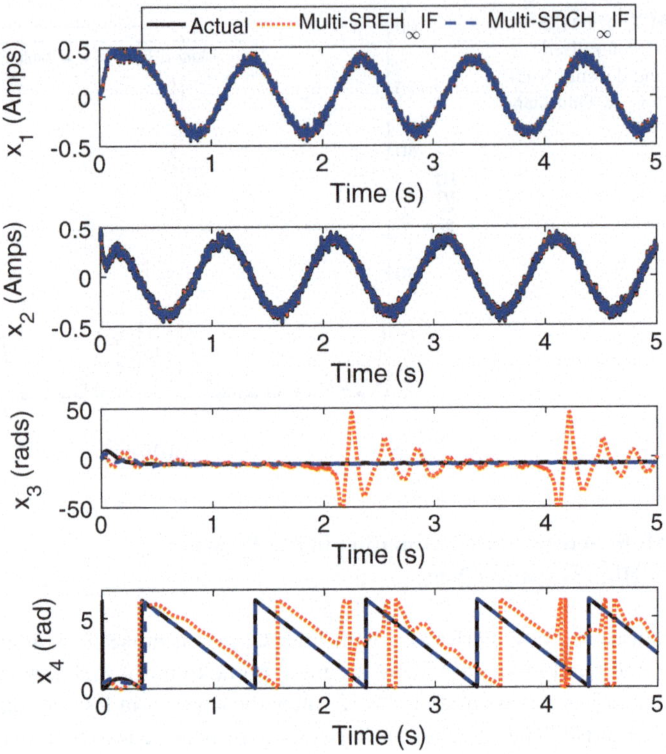

Fig. 6.10 PMSM states using decentralised SRE$\mathcal{H}_\infty$IF and decentralised SRC$\mathcal{H}_\infty$IF for high Gaussian noise

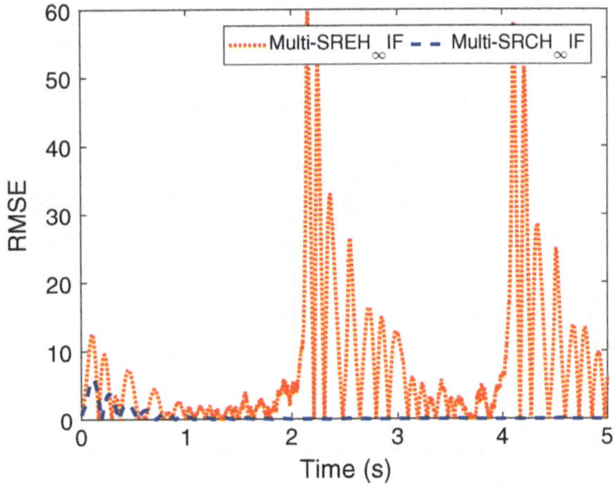

Fig. 6.11 RMSE of PMSM speed using decentralised SRE$\mathcal{H}_\infty$IF and decentralised SRC$\mathcal{H}_\infty$IF for high Gaussian noise

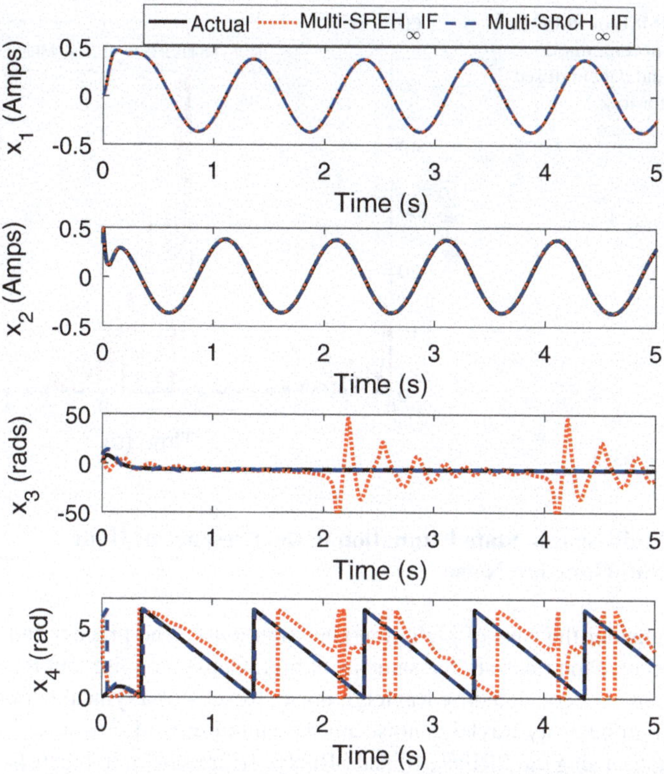

Fig. 6.12 PMSM states using decentralised SRE$\mathcal{H}_\infty$IF and decentralised SRC$\mathcal{H}_\infty$IF for low non-Gaussian noise

(2009). In computer simulations, Rayleigh noise can be generated using the MATLAB command 'raylrnd'. To demonstrate the efficacy of the state estimators in the presence of non-Gaussian noise, simulations are performed for low and high intensity Rayleigh noises.

First we consider the case of low non-Gaussian noise. The process and measurement covariance matrices are the same as of low Gaussian noise discussed earlier, however, the plant and measurements are corrupted by Rayleigh noise. Results of a typical simulation run using the Rayleigh noise are shown in Fig. 6.12. The actual states and their estimates using the SRE$\mathcal{H}_\infty$IF and SRC$\mathcal{H}_\infty$IF are shown in Fig. 6.12. In this case also the first two states and their estimates using SRE$\mathcal{H}_\infty$IF and SRC$\mathcal{H}_\infty$IF almost overlap and estimation errors of speed and rotor position using the SRE$\mathcal{H}_\infty$IF are larger. The RMSEs for the speed using SRE$\mathcal{H}_\infty$IF and SRC$\mathcal{H}_\infty$IF are shown in Fig. 6.13. The maximum RMSE error is 39.74 for the SRE$\mathcal{H}_\infty$IF and 8.24 for SRC$\mathcal{H}_\infty$IF, and the average RMSE error is 12.89 for the SRE$\mathcal{H}_\infty$IF and 1.64 for SRC$\mathcal{H}_\infty$IF.

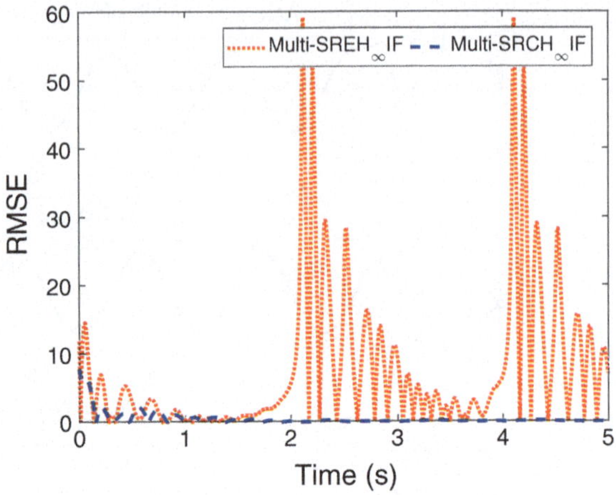

Fig. 6.13 RMSE of PMSM speed using decentralised SRE$\mathcal{H}_\infty$IF and decentralised SRC$\mathcal{H}_\infty$IF for low non-Gaussian noise

6.5.4.4 Multi-sensor State Estimation in the Presence of High Non-Gaussian Noise

We now consider the case of high, non-Gaussian noise. The process and measurement covariance matrices are the same as of high Gaussian noise, but the plant and measurements are corrupted by Rayleigh noise. Results of a typical simulation run using the high intensity Rayleigh noise are shown in Fig. 6.14. The actual states and their estimates using the SRE$\mathcal{H}_\infty$IF and SRC$\mathcal{H}_\infty$IF are shown in Fig. 6.14. The first two states are seen to be noisy due to the high intensity Rayleigh noise added to them. In this case also the first two states and their estimates using SRE$\mathcal{H}_\infty$IF and SRC$\mathcal{H}_\infty$IF almost overlap and estimation errors of speed and rotor position using the SRE$\mathcal{H}_\infty$IF are larger. The RMSEs for the speed using SRE$\mathcal{H}_\infty$IF and SRC$\mathcal{H}_\infty$IF are shown in Fig. 6.15. The maximum RMSE error is 48.37 for the SRE$\mathcal{H}_\infty$IF and 32 for SRC$\mathcal{H}_\infty$IF, and the average RMSE error is 12.82 for the SRE$\mathcal{H}_\infty$IF and 5.52 for SRC$\mathcal{H}_\infty$IF.

6.5.4.5 Multi-sensor State Estimation in the Presence of Near-Perfect Measurements

This part presents the simulation results with near-perfect plant and measurement models in the presence of Gaussian and non-Gaussian noise. It is assumed that the sensors provide near-perfect measurements and the process models are also almost perfect. The process and measurement noise covariance matrices are chosen as

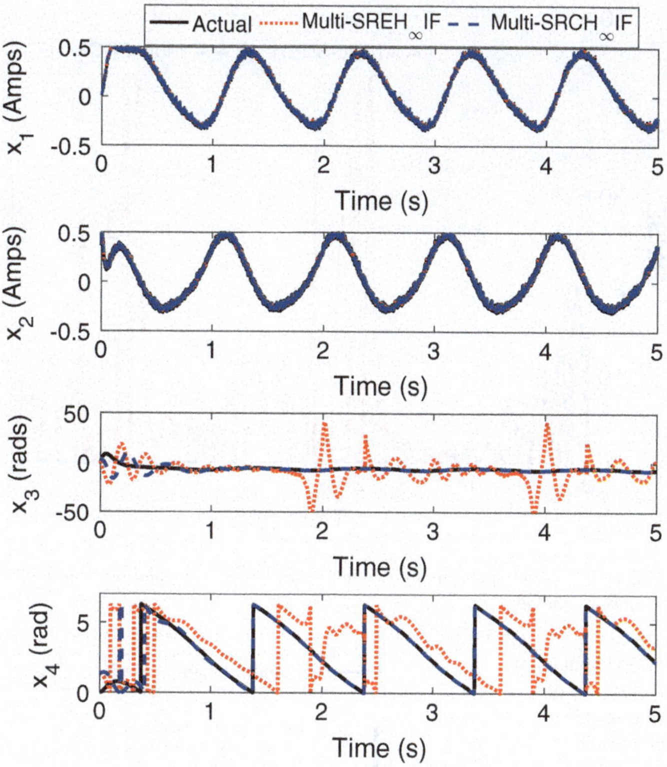

Fig. 6.14 PMSM states using decentralised SRE$\mathcal{H}_\infty$IF and decentralised SRC$\mathcal{H}_\infty$IF for high non-Gaussian noise

$$\mathbf{Q} = \begin{bmatrix} 1\times10^{-20} & 0 & 0 & 0 \\ 0 & 1\times10^{-20} & 0 & 0 \\ 0 & 0 & 1\times10^{-20} & 0 \\ 0 & 0 & 0 & 1\times10^{-20} \end{bmatrix}, \mathbf{R}_1 = \begin{bmatrix} 1\times10^{-20} & 0 \\ 0 & 1\times10^{-20} \end{bmatrix},$$

$$\mathbf{R}_2 = \begin{bmatrix} 1\times10^{-20} & 0 \\ 0 & 1\times10^{-20} \end{bmatrix}.$$

The square-root filters have inherent tendency to effectively estimate the states in the presence of near-perfect measurements, for more details please see Tsyganova and Kulikova (2013), Grewal and Andrews (2010), Simon (2006), Chandra (2014). The estimated and actual states with near-perfect measurements are very close for both the Gaussian and Rayleigh noises, and hence the plots of states and their estimates are not shown. The root means square errors (RMSEs) for the speed in the presence of Gaussian and non-Gaussian noise using multi-sensor SRC$\mathcal{H}_\infty$IF are shown in Fig. 6.16. The RMSE errors for both cases are very low. The maximum RMSE error is 3.18 for the Gaussian noise and 3.19 for the Rayleigh noise, and the average

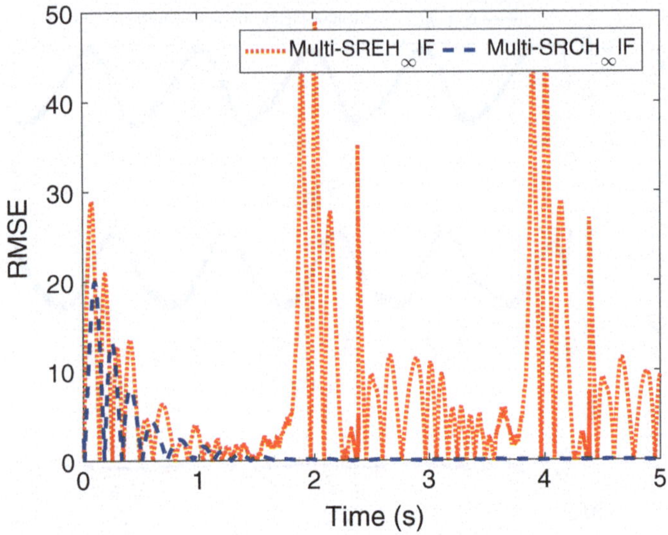

Fig. 6.15 RMSE of PMSM speed using decentralised SRE$\mathcal{H}_\infty$IF and decentralised SRC$\mathcal{H}_\infty$IF for high non-Gaussian noise

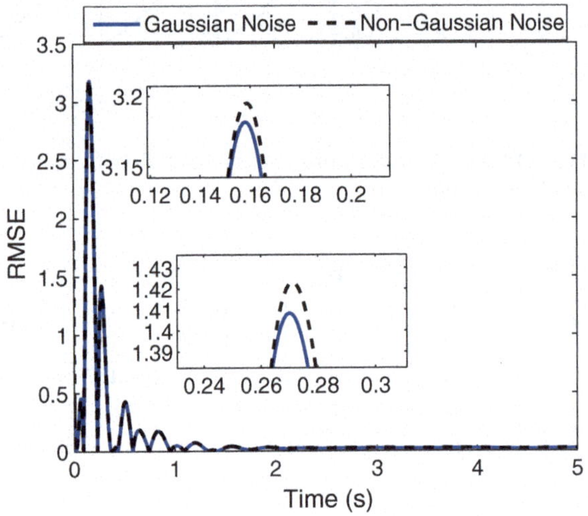

Fig. 6.16 PMSM simulation results in the presence of near-perfect Gaussian and non-Gaussian noise in multi-sensor SRC$\mathcal{H}_\infty$IF

RMSE error is 0.5816 for the Gaussian noise and 0.5869 for the non-Gaussian noise respectively.

Since the square-root extended $\mathcal{H}_\infty$ information filter is a Jacobian based approach, it is sometimes difficult to implement on various applications for example when the Jacobians are ill-conditioned or for piecewise nonlinear systems. However, the square-root cubature $\mathcal{H}_\infty$ information filter is a Jacobian-free approach and hence one can avoid the difficulties due to Jacobians. The simulation results in this numerical example show the efficacy of SRC$\mathcal{H}_\infty$IF. It can be observed that this type of filter has a tendency to handle low and high Gaussian noises and non-Gaussian noises as well. Further, the square-root cubature $\mathcal{H}_\infty$ information filter can be very useful in applications where the sensor measurements are near-perfect. This shows that the square root cubature $hinf$ information filter has certain *robustness*, the property will be touched upon in the next chapter.

The simulations have been performed using 64 bit Matlab on a computer with the processor speed on 3.4 GHz. The average computation time, for the simulations included in this example, are 3.1 and 3.6 s for SRE$\mathcal{H}_\infty$IF and SRC$\mathcal{H}_\infty$IF, respectively. However, it can be seen that in all the simulations the quality of the state estimate using SRC$\mathcal{H}_\infty$IF is far better than SRE$\mathcal{H}_\infty$IF.

6.6 State Estimation of a Continuous Stirred Tank Reactor (CSTR)

In this section, a continuous stirred tank reactor (CSTR) system is to be considered. The state of CSTR will be estimated in the presence of Gaussian and non-Gaussian noises, in order to evaluate the performance of the square-root cubature $\mathcal{H}_\infty$ filter (SRC$\mathcal{H}_\infty$F) in comparison with the performance of the square-root cubature Kalman filter (SRCKF).

6.6.1 The Model

The process model for an irreversible, first-order chemical reaction, $A \rightarrow B$ which occurs in the continuous stirred tank reactor (CSTR) is as follows Uppal et al. (1974), Kurtz and Henson (1995):

$$\dot{C_A} = \frac{q}{V}(C_{Af} - C_A) - k_0 \exp\left(\frac{-E}{RT}\right) C_A \tag{6.179}$$

$$\dot{T} = \frac{q}{V}(T_f - T) + \frac{-\nabla H}{\rho C_p} k_0 \exp\left(\frac{-E}{RT}\right) C_A + \frac{UA}{V\rho C_p}(T_c - T) \tag{6.180}$$

The state variables are the concentration, C_A, and the temperature of the reactor, T; the measured output, h, depends on the temperature T.

The following parameters are considered in the simulations Kurtz and Henson (1995):
$q = 100$ L/min, $\frac{E}{R} = 8750$ K, $C_{Af} = 1$ mol/L, $K_0 = 7.5 \times 10^{10}\text{min}^{-1}$, $T_f = 350$ K, $UA = 5000$ J/minK, $V = 100$ L, $T_c = 300$ K, $\rho = 1000$ g/L, $C_A = 0.5$ mol/L, $C_p = 0.239$ J/gK, $T = 350$ K and $(\nabla H) = 5000$ J/mol.

The process and measurement noises, w_k and v_k, are added to the process and measurement models, respectively. The control variable, the coolant temperature T_c, is computed using the input–output feedback linearisation Kurtz and Henson (1995)

$$T_c = \frac{\zeta - L_f h(x)}{L_g h(x)} \tag{6.181}$$

The Lie derivatives L_f and L_g, and ζ are

$$L_f h(x) = \frac{q}{V}(T_f - T) + \frac{-\nabla H}{\rho C_p} k_0 \exp\left(\frac{-E}{RT}\right) C_A - \frac{UA}{V\rho C_p}(T) \tag{6.182}$$

$$L_g h(x) = \frac{UA}{V\rho C_p} \tag{6.183}$$

$$\zeta = 50z + 10(T_{setpoint} - T) \tag{6.184}$$

where z can be obtained by integrating the following:

$$\dot{z} = T_{setpoint} - T.$$

The control input is generated based on the estimated states. The plant is discretised using Euler's method with a sampling time of 0.01s and the reactor setpoint is 400 K. The objective is to estimate the full state vector of the closed-loop CSTR using the noisy temperature measurements. The simulations with near-perfect measurement noise are performed. The selected tuning parameters for the ***square-root cubature H_∞ filter*** are $\gamma = 1$ and

$$Q = \begin{bmatrix} 0.001 & 0 \\ 0 & 0.5 \end{bmatrix} \quad and \quad R = 0.001$$

6.6.2 *State Estimation in the Presence of Non-Gaussian Noises*

A square-root cubature $\mathcal{H}_\infty$ filter is used to estimate the states of the CSTR in the presence of non-Gaussian, non-zero mean, time-varying noises. The process and measurement noises are

$$w_k = \begin{bmatrix} 0.01 + 0.05 \times sin(0.1k) & 0 \\ 0 & 1 + sin(0.1k) \end{bmatrix} \tag{6.185}$$

$$v_k = 1 + 0.1 \times sin(0.1k). \tag{6.186}$$

The CSTR model is initialised at $\mathbf{x}_0 = [0.4 \quad 340]^T$ and the chosen associated auxiliary matrix is $\mathbf{P}_{0|0} = [0.001 \quad 0; 0 \quad 0.01]$. The initial state for estimation is ran-

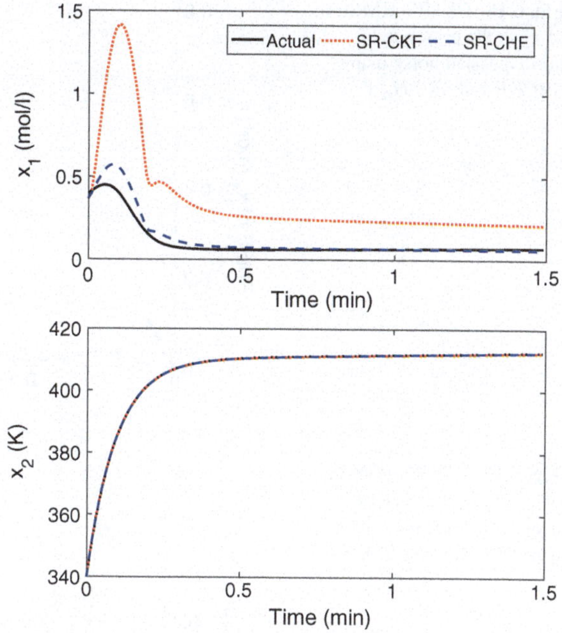

Fig. 6.17 Actual and estimated states of CSTR for non-Gaussian noise using SRCKF and SRC$\mathcal{H}_\infty$F

domly selected from a uniform distribution of $\mathbb{U}(\mathbf{x}_0, \mathrm{b}f\, P_{0|0})$. The maximum magnitudes of w_k and v_k are chosen in the noise covariance matrices for the square-root CKF. The actual and estimated states, and the root mean square errors (RMSEs) using the square-root CKF and square root cubature $\mathcal{H}_\infty$ filter are shown in Figs. 6.17 and 6.18, respectively. Please note that in some simulation plots the curves generated from using SRC$\mathcal{H}_\infty$F could be denoted by SRCHF. The offsets in the RMSE plots are due to the non-zero bias of the noises, which were intentionally added to verify the effectiveness of the square root cubature $\mathcal{H}_\infty$ filter in the presence of non-zero mean noises. The maximum RMSEs for the square-root CKF and the square root cubature $\mathcal{H}_\infty$ filter are 0.7372 and 0.1116, respectively. Hence, the square-root cubature $\mathcal{H}_\infty$ filter appears to be better suited for the nonlinear state estimation in the presence of non-Gaussian noises. The filters' response in the presence of Gaussian noises has not been compared as Kalman filter's performance can be achieved by tuning the $\mathcal{H}_\infty$ filter parameters. That is, the square-root CKF's performance could be replicated by using the square-root cubature $\mathcal{H}_\infty$ filter, but the reverse is not always possible.

6.6.3 *State Estimation with Near-Perfect Measurements*

To illustrate the effectiveness of the square-root cubature $\mathcal{H}_\infty$ filter, the case of near-perfect measurements has been considered. Non-square-root estimators have the tendency to diverge due to singularity issues. There are many ad-hoc methods

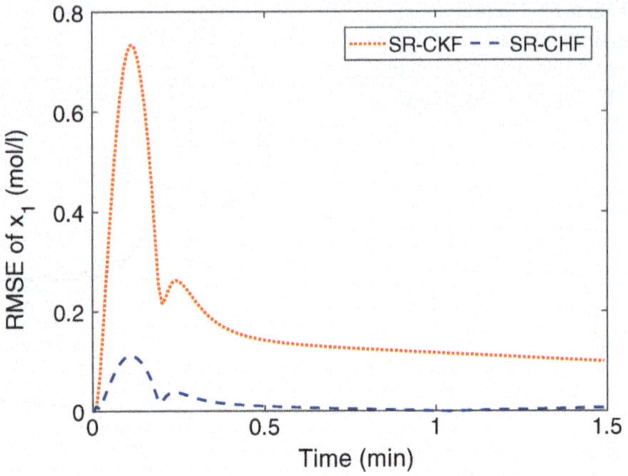

Fig. 6.18 CSTR's RMSEs of x_1 in the presence of non-Gaussian noise using SRCKF and SRC$\mathcal{H}_\infty$F

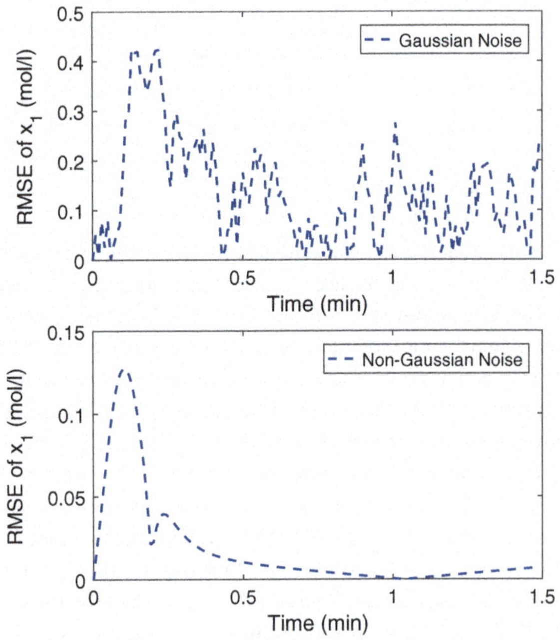

Fig. 6.19 CSTR's RMSEs of x_1 in the presence of near-perfect measurement using SRCKF and SRC$\mathcal{H}_\infty$F

available to circumvent these problems. However, in most cases, the square-root filter inherently solves these numerical issues (Anderson and Moore 1979; Grewal and Andrews 2001; Simon 2006). Simulations using the *non-square root* cubature $\mathcal{H}_\infty$ filter for the perfect case are not presented, as its response diverges after 3 4 time steps. In the non-Gaussian noise case, the simulations in the previous part are repeated with near-perfect measurements. In the Gaussian noise case, the standard deviations of the two states for process noise are considered as 0.1. The magnitude of the measurement noise is assumed to be zero (near-perfect measurement) for both

Gaussian and non-Gaussian simulations. The corresponding results are shown in Fig. 6.19. It can be seen that the RMSEs, using square-root cubature $\mathcal{H}_\infty$ filters in the presence of Gaussian and non-Gaussian, are within an acceptable range.

6.7 Summary

In this chapter a few variants of cubature Kalman filters (CKF) have been discussed, namely the cubature information filters (CIF), the cubature $\mathcal{H}_\infty$ filters (C$\mathcal{H}_\infty$F) and the cubature $\mathcal{H}_\infty$ information filters (C$\mathcal{H}_\infty$IF). Cubature information filters are derived in the information domain where the information vector and information matrices are propagated. The key advantage of CIF is computationally simple during the update stage. Further, CIF can be conveniently extended to systems with multiple sensors. In the other direction, the CKF has been explored for non-Gaussian systems, where CKF is fused with $\mathcal{H}_\infty$ filter to form C$\mathcal{H}_\infty$F. Finally, C$\mathcal{H}_\infty$F has been fused with the information filter to form C$\mathcal{H}_\infty$IF which is suitable for the multi-sensor state estimation. For numerical accuracy, square-root versions of these filters have also been derived and discussed. To show the efficiency and effectiveness of various approaches, detailed studies on two examples, a permanent magnet synchronous motor (PMSM) system and a continuous stirred tank reactor (CSTR) system have been included in this chapter. Hopefully, those two example cases demonstrated the usage and characteristics of those nonlinear state estimation approaches.

References

Anderson BDO, Moore JB (1979) Optimal filtering, vol 21. Prentice-Hall, Englewood Cliffs, pp 22–95

Arasaratnam I, Haykin S (2009) Cubature kalman filters. IEEE Trans Autom control 54(6):1254–1269

Bar-Shalom Y, Li XR, Kirubarajan T (2004) Estimation with applications to tracking and navigation: theory algorithms and software. Wiley, New York

Chandra KB, Gu DW, Postlethwaite I (2013) Square root cubature information filter. IEEE SensJ 13(2):750–758

Chandra KPB, Gu D-W, Postlethwaite I (2014) A cubature H_∞ filter and its square-root version. Int J Control 87(4):764–776

Chandra KPB, Gu D-W, Postlethwaite I (2016) Cubature H_∞ information filter and its extensions. Eur J Control 29:17–32

Eustice RM, Singh H, Leonard JJ, Walter MR (2006) Visually mapping the rms titanic: conservative covariance estimates for slam information filters. Int J Robot Res 25(12):1223–1242

Grewal M, Andrews A (2001) Kalman filtering: theory and practice using matlab

Grewal MS, Andrews AP (2010) Applications of kalman filtering in aerospace 1960 to the present [historical perspectives]. IEEE Control Syst Mag 30(3):69–78

Hassibi B, Kailath T, Sayed AH (2000) Array algorithms for h/sup/spl infin//estimation. IEEE Trans Autom Control 45(4):702–706

Ishihara J, Macchiavello B, Terra M (2006) H8 estimation and array algorithms for discrete-time descriptor systems. In: 2006 45th IEEE conference on decision and control. IEEE, pp 4740–4745
Kailath T, Sayed AH, Hassibi B (2000) Linear estimation, vol 1. Prentice Hall, Upper Saddle River
Kurtz MJ, Henson MA (1995) Nonlinear output feedback control of chemical reactors. In: Proceedings of the 1995 American control conference, vol 4 IEEE, pp 2667–2671
Lee D-J (2008) Nonlinear estimation and multiple sensor fusion using unscented information filtering. IEEE Signal Process Lett 15:861–864
Li W, Jia Y (2010) H-infinity filtering for a class of nonlinear discrete-time systems based on unscented transform. Signal Process 90(12):3301–3307
Mutambara AG (1998) Decentralized estimation and control for multisensor systems. CRC Press, Boca Raton
Park P, Kailath T (1995) New square-root algorithms for kalman filtering. IEEE Trans Autom Control 40(5):895–899
Raol J, Girija G (2002) Sensor data fusion algorithms using square-root information filtering. IEE Proc-Radar Sonar Navig 149(2):89–96
Rigatos GG (2009) Particle filtering for state estimation in nonlinear industrial systems. IEEE Trans Instrum Meas. 58(11):3885–3900
Shen X-M, Deng L (1997) Game theory approach to discrete h_∞ filter design. IEEE Trans Signal Process 45(4):1092–1095
Sibley G, Sukhatme GS, Matthies LH (2006) The iterated sigma point kalman filter with applications to long range stereo. Robot: Sci Syst 8:235–244
Simon D (2006) Optimal state estimation: Kalman, H infinity, and nonlinear approaches. Wiley, New York
Terra MH, Ishihara JY, Jesus G (2009) Fast array algorithms for h_∞ information estimation of rectangular discrete-time descriptor systems. In: 2009 IEEE control applications, (CCA) & intelligent control, (ISIC). IEEE, pp 637–642
Tsyganova J, Kulikova M (2013) State sensitivity evaluation within ud based array covariance filters. IEEE Trans Autom Control 58(11):2944–2950
Uppal A, Ray W, Poore A (1974) On the dynamic behavior of continuous stirred tank reactors. Chem Eng Sci 29(4):967–985

Chapter 7
More Estimation Methods and Beyond

7.1 Introductory Remarks

This chapter presents two more nonlinear state estimation methods: the unscented Kalman filter (UKF) and the state-dependent Riccati equation (SDRE) observer. Though the UKF filter and SDRE observer are based on different philosophies, both of them are derivative-free, nonlinear, state estimators. Their formulations and applications to nonlinear systems are described in this chapter with illustrative application examples. Towards the end of this chapter, consideration is to be given to the *robustness* of state estimators in general. Uncertainty is always a crucial issue in control systems design and operations, and so is with the state estimation. After a very brief description of system dynamics perturbation and signal disturbances, the scenario of observation data missing occurring in the state estimation will be discussed. The method of linear prediction (LP) is particularly introduced to alleviate the adverse effect of data missing in the state estimation, together with a numerical example.

7.2 Unscented Kalman Filter

In Sect. 4.6, effects of the Jacobian linearisation in calculating the mean and covariance of a nonlinear transformation was analysed. In this section, a so-called unscented transformation and the corresponding unscented Kalman filter (UKF) will be discussed. The unscented transformation is derivative free and is founded on the intuition that "*it is easier to approximate a Gaussian distribution than it is to approximate an arbitrary nonlinear function*" (Julier and Uhlmann 2004). In the unscented transformation, a set of deterministic sigma points are chosen and are propagated through the nonlinear function. Further, a weighted mean and covariance are then evaluated. The unscented transform ensures the higher accuracy than a linearisation approach.

K. P. B. Chandra and D.-W. Gu, *Nonlinear Filtering*,
https://doi.org/10.1007/978-3-030-01797-2_7

7.2.1 Unscented Transformation

Consider the nonlinear function given by $\mathbf{s} = \mathbf{h}(\mathbf{x})$ where $\mathbf{x} \in \mathbb{R}^n$. The mean and covariance of the random variable $\mathbf{x}$ are $\bar{\mathbf{x}}$ and $\mathbf{P}_\mathbf{x}$, respectively. The mean and covariance of $\mathbf{s}$, $\bar{\mathbf{s}}$ and $\mathbf{P}_s$, can be obtained by using the unscented transformation described in Algorithm 7.1.

Polar to Cartesian Coordinate Transformation—Unscented Transformation

Consider the polar to cartesian nonlinear transformation given by

---

Algorithm 7.1 Unscented Transform

1: Compute the $2n + 1$ weighted sigma points

$$\begin{aligned} \chi_0 &= \bar{\mathbf{x}} \\ \chi_i &= \bar{\mathbf{x}} + \left[\sqrt{(n+\lambda)\mathbf{P}_\mathbf{x}}\right]_i, \quad i = 1, \ldots, n \\ \chi_i &= \bar{\mathbf{x}} - \left[\sqrt{(n+\lambda)\mathbf{P}_\mathbf{x}}\right]_i, \quad i = n+1, \ldots, 2n \end{aligned} \tag{7.1}$$

where, $[\ldots]_i$ denotes the ith column of $[\ldots]$. Set the corresponding weights as

$$\begin{aligned} W_0^\mathbf{x} &= \frac{\lambda}{n+\lambda} \\ W_0^P &= \frac{\lambda}{n+\lambda} + (1 - \alpha^2 + \beta) \\ W_i^\mathbf{x} &= \frac{1}{2(n+\lambda)}, \quad i = 1, \ldots, 2n \\ W_i^P &= W_i^\mathbf{x}, \quad i = 1, \ldots, 2n \end{aligned} \tag{7.2}$$

where

$$\lambda = \alpha^2(n + \kappa) - n. \tag{7.3}$$

The suggested values for α, β and κ are $1 \times 10^{-2} or 1 \times 10^{-3}$, 2 and $3 - n$, respectively (Van Der Merwe 2004; Julier et al. 2000).

2: Propagate the sigma points through the nonlinear function

$$\mathfrak{s}_i = \mathbf{h}(\chi_i), \quad i = 0, \ldots, 2n. \tag{7.4}$$

3: The mean and covariance of $\mathbf{s}$ are

$$\bar{\mathfrak{s}} \approx \sum_{i=0}^{2n} W_i^\mathbf{x} \mathfrak{s}_i, \quad i = 0, \ldots, 2n \tag{7.5}$$

$$\mathbf{P}_s \approx \sum_{i=0}^{2n} W_i^P (\mathfrak{s}_i - \bar{\mathfrak{s}})(\mathfrak{s}_i - \bar{\mathfrak{s}})^T, \quad i = 0, \ldots, 2n.. \tag{7.6}$$

$$\begin{bmatrix} x \\ y \end{bmatrix} = \mathbf{h}(\mathbf{x}) = \begin{bmatrix} r\cos\theta \\ r\sin\theta \end{bmatrix} \tag{7.7}$$

where $\mathbf{x}$, consisting of the range and bearing of the target, is $\mathbf{x} = [r \quad \theta]^T$, and $[x \quad y]^T$ are the cartesian coordinates of the target. This transformation is important in real world applications such as the simultaneous localisation and mapping (SLAM) which has been discussed earlier in the book. It is assumed that r and θ are two independent variables with means $\bar{r}$ and $\bar{\theta}$, and the corresponding standard deviations are σ_r and σ_θ, respectively. In this section, the unscented transformation given in Algorithm 7.1 will be used to obtain the mean and covariance of Eq. (5.25). The size of the state vector is $n = 2$, the parameters α, β and κ are selected as 0.01, 2 and 1, respectively, and λ using Eq. (7.14) is -1.9997. The remaining parameters used in the simulations are the same as given in Sect. 4.6.

The sigma points using Eq. (7.1) can be calculated as

$$\begin{aligned}
\boldsymbol{\chi}_0 &= \begin{bmatrix} \bar{r} \\ \bar{\theta} \end{bmatrix} = \begin{bmatrix} 1 \\ \frac{\pi}{2} \end{bmatrix} \\
\boldsymbol{\chi}_1 &= \bar{\mathbf{x}} + \left[\sqrt{(n+\lambda)\mathbf{P}_\mathbf{x}}\right]_1 = \begin{bmatrix} 1 + \sigma_r\sqrt{n+\lambda} \\ \frac{\pi}{2} \end{bmatrix} \\
\boldsymbol{\chi}_2 &= \bar{\mathbf{x}} + \left[\sqrt{(n+\lambda)\mathbf{P}_\mathbf{x}}\right]_2 = \begin{bmatrix} 1 \\ \frac{\pi}{2} + \sigma_\theta\sqrt{n+\lambda} \end{bmatrix} \\
\boldsymbol{\chi}_3 &= \bar{\mathbf{x}} - \left[\sqrt{(n+\lambda)\mathbf{P}_\mathbf{x}}\right]_3 = \begin{bmatrix} 1 - \sigma_r\sqrt{n+\lambda} \\ \frac{\pi}{2} \end{bmatrix} \\
\boldsymbol{\chi}_4 &= \bar{\mathbf{x}} - \left[\sqrt{(n+\lambda)\mathbf{P}_\mathbf{x}}\right]_4 = \begin{bmatrix} 1 \\ \frac{\pi}{2} - \sigma_\theta\sqrt{n+\lambda} \end{bmatrix}
\end{aligned} \tag{7.8}$$

where,

$$\mathbf{P}_\mathbf{x} = \begin{bmatrix} \sigma_r^2 & 0 \\ 0 & \sigma_\theta^2 \end{bmatrix} \tag{7.9}$$

The corresponding weights are

$$W_0^\mathbf{x} = \frac{\lambda}{n+\lambda} = -6665.66 \tag{7.10}$$

$$W_0^P = \frac{\lambda}{n+\lambda} + (1 - \alpha^2 + \beta) = -6662.66 \tag{7.11}$$

$$W_1^\mathbf{x} = W_2^\mathbf{x} = W_3^\mathbf{x} = W_4^\mathbf{x} = 1666 \tag{7.12}$$

$$W_1^P = W_2^P = W_3^P = W_4^P = 1666 \tag{7.13}$$

where

$$\lambda = \alpha^2(n+\kappa) - n = -1.9997. \tag{7.14}$$

The transformed sigma points using Eq. (7.4) are

$$\begin{aligned}
\mathfrak{s}_0 &= \mathbf{h}(\chi_0) = \mathbf{h}\left(\begin{bmatrix}\bar{r}\\ \bar{\theta}\end{bmatrix}\right) = \begin{bmatrix}0\\1\end{bmatrix}\\
\mathfrak{s}_1 &= \mathbf{h}(\chi_1) = \mathbf{h}\left(\begin{bmatrix}1+\sigma_r\sqrt{n+\lambda}\\ \frac{\pi}{2}\end{bmatrix}\right) = \begin{bmatrix}0\\ 1+\sigma_r\sqrt{(n+\lambda)}\end{bmatrix}\\
\mathfrak{s}_2 &= \mathbf{h}(\chi_2) = \mathbf{h}\left(\begin{bmatrix}1\\ \frac{\pi}{2}+\sigma_\theta\sqrt{n+\lambda}\end{bmatrix}\right) = \begin{bmatrix}\cos\left(\frac{\pi}{2}+\sigma_\theta\sqrt{(n+\lambda)}\right)\\ \sin\left(\frac{\pi}{2}+\sigma_\theta\sqrt{(n+\lambda)}\right)\end{bmatrix}\\
\mathfrak{s}_3 &= \mathbf{h}(\chi_3) = \mathbf{h}\left(\begin{bmatrix}1-\sigma_r\sqrt{n+\lambda}\\ \frac{\pi}{2}\end{bmatrix}\right) = \begin{bmatrix}0\\ 1-\sigma_r\sqrt{(n+\lambda)}\end{bmatrix}\\
\mathfrak{s}_4 &= \mathbf{h}(\chi_4) = \mathbf{h}\left(\begin{bmatrix}1\\ \frac{\pi}{2}-\sigma_\theta\sqrt{n+\lambda}\end{bmatrix}\right) = \begin{bmatrix}\cos\left(\frac{\pi}{2}-\sigma_\theta\sqrt{(n+\lambda)}\right)\\ \sin\left(\frac{\pi}{2}-\sigma_\theta\sqrt{(n+\lambda)}\right)\end{bmatrix}.
\end{aligned} \tag{7.15}$$

Once the sigma points and its corresponding weights, and the transformed sigma points, have been obtained, the mean and covariance of $\mathbf{s}$ can be calculated as

$$\bar{\mathfrak{s}} \approx \sum_{i=0}^{4} W_i^{\mathbf{x}}\mathfrak{s}_i \tag{7.16}$$

$$\mathbf{P}_s \approx \sum_{i=0}^{4} W_i^{P}(\mathfrak{s}_i - \bar{\mathfrak{s}})(\mathfrak{s}_i - \bar{\mathfrak{s}})^T. \tag{7.17}$$

A similar simulation scenario as that given in Sect. 4.6 is repeated with the unscented transformation and the corresponding results are shown in Fig. 7.1. The mean and covariance of the unscented transformation along with the true and linearised ones are shown in Fig. 7.1. It is very hard to see the true mean as it is hidden behind the unscented transform. The true, linearised and unscented transformation means are located at (0, 0.9798), (0, 1) and (0, 0.9797), respectively. The error covariances for true, linearised and unscented transformation are

$$\begin{bmatrix}0.1991 & 0\\ 0 & 0.0213\end{bmatrix}, \begin{bmatrix}0.2015 & 0\\ 0 & 0.0115\end{bmatrix} \text{and} \begin{bmatrix}0.2015 & 0\\ 0 & 0.0310\end{bmatrix}. \tag{7.18}$$

The uncertainty ellipse using the unscented transformation is comparatively unbiased as compared to that of the linearised uncertainty ellipse. It can also be seen that the uncertainty ellipse for the unscented transformation given in Fig. 7.1 does not match with that of the true ellipse along the y-axis. One of the reasons for this mismatch is due to the negative λ. The unscented transform response can be further improved by tuning α, β and κ.

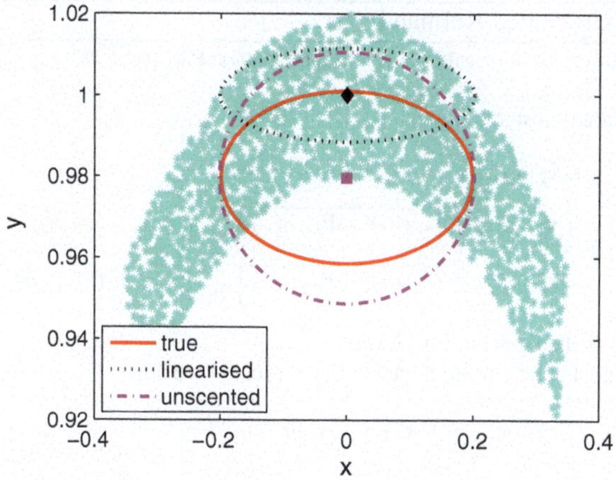

Fig. 7.1 2000 random measurements are generated with range and bearings, which are uniformly distributed between ±0.02 and ±20°. These random points are then processed through the non-linear polar to cartesian coordinate transformation and are shown as ∗. The true, linearised and unscented transformation means are represented by •, ♦ and ■, respectively. True, linearised and unscented transformation uncertainty ellipses are represented by solid, dotted and dashed-dotted lines, respectively

7.2.2 *Unscented Kalman Filter*

The unscented Kalman filter (UKF) is a recursive filter based on the unscented transformation. In Sect. 7.2.1, the advantages of the unscented transformation over the linearisation approximation were demonstrated. The unscented transformation is the backbone of the UKF.

Consider the discrete-time process and measurement models given by

$$\mathbf{x}_k = \mathbf{f}(\mathbf{x}_{k-1}, \mathbf{u}_{k-1}) + \mathbf{w}_{k-1} \tag{7.19}$$

$$\mathbf{z}_k = \mathbf{h}(\mathbf{x}_k) + \mathbf{v}_k, \tag{7.20}$$

.

Similar to the extended Kalman filter (EKF) and many other filters, UKF can also be expressed in two stages, prediction and measurement update stages, and is briefed in Algorithm 7.2. For more details on UKF, please see, for example Simon (2006) and Julier et al. (2000).

Algorithm 7.2 Unscented Kalman Filter

Initialise the state vector, $\hat{\mathbf{x}}_{0|0}$, and the covariance matrix, $\mathbf{P}_{0|0}$ (set $k = 1$)

Prediction

Calculate the prediction sigma points

$$\begin{aligned}
\chi_{0,k-1|k-1} &= \hat{\mathbf{x}}_{k-1|k-1} \\
\chi_{i,k-1|k-1} &= \hat{\mathbf{x}}_{k-1|k-1} + \left[\sqrt{(n+\lambda)\mathbf{P}_{k-1|k-1}}\right]_i, \quad i = 1, \ldots, n \\
\chi_{i,k-1|k-1} &= \hat{\mathbf{x}}_{k-1|k-1} - \left[\sqrt{(n+\lambda)\mathbf{P}_{k-1|k-1}}\right]_i, \quad i = n+1, \ldots, 2n
\end{aligned} \tag{7.21}$$

where λ can be calculated using Eq. (7.3).

Propagate the sigma points through the nonlinear process model

$$\chi^*_{i,k|k-1} = \mathbf{f}(\chi_{i,k-1|k-1}, \mathbf{u}_{k-1}), \quad i = 0, \ldots, 2n. \tag{7.22}$$

Predicted state and covariance can be obtained as

$$\hat{\mathbf{x}}_{k|k-1} = \sum_{i=0}^{2n} W_i^x \chi^*_{i,k|k-1} \tag{7.23}$$

$$\mathbf{P}_{k|k-1} = \sum_{i=0}^{2n} W_i^P (\chi^*_{i,k|k-1} - \hat{\mathbf{x}}_{k|k-1})(\chi^*_{i,k|k-1} - \hat{\mathbf{x}}_{k|k-1})^T + \mathbf{Q}_{k-1} \tag{7.24}$$

where the weights, W_i^x and W_i^P, can be calculated using Eq. (7.2).

Measurement update

Calculate the update sigma points (these sigma points are calculated using predicted mean and covariance, $\hat{\mathbf{x}}_{k|k-1}$ and $\mathbf{P}_{k|k-1}$)

$$\begin{aligned}
\chi_{0,k|k-1} &= \hat{\mathbf{x}}_{k|k-1} \\
\chi_{i,k|k-1} &= \hat{\mathbf{x}}_{k|k-1} + \left[\sqrt{(n+\lambda)\mathbf{P}_{k|k-1}}\right]_i, \quad i = 1, \ldots, n \\
\chi_{i,k|k-1} &= \hat{\mathbf{x}}_{k|k-1} - \left[\sqrt{(n+\lambda)\mathbf{P}_{k|k-1}}\right]_i, \quad i = n+1, \ldots, 2n
\end{aligned} \tag{7.25}$$

Propagate the sigma points through the nonlinear measurement model

$$\mathbf{z}_{i,k|k-1} = \mathbf{h}(\chi_{i,k|k-1}, \mathbf{u}_k), \quad i = 0, \ldots, 2n. \tag{7.26}$$

Predicted measurement, covariance and cross-covariance can be calculated as

$$\hat{\mathbf{z}}_{k|k-1} = \sum_{i=0}^{2n} W_i^x \mathbf{z}_{i,k|k-1} \tag{7.27}$$

$$\mathbf{P}_{zz,k|k-1} = \sum_{i=0}^{2n} W_i^P (\mathbf{z}_{i,k|k-1} - \hat{\mathbf{z}}_{k|k-1})(\mathbf{z}_{i,k|k-1} - \hat{\mathbf{z}}_{k|k-1})^T + \mathbf{R}_k \tag{7.28}$$

$$\mathbf{P}_{xz,k|k-1} = \sum_{i=0}^{2n} W_i^P (\chi_{i,k|k-1} - \hat{\mathbf{x}}_{k|k-1})(\mathbf{z}_{i,k|k-1} - \hat{\mathbf{z}}_{k|k-1})^T \tag{7.29}$$

where the weights, W_i^x and W_i^P, can be calculated using Eq. (7.2).

Update mean and error covariance can be obtained as

$$\hat{\mathbf{x}}_{k|k} = \hat{\mathbf{x}}_{k|k-1} + \mathbf{K}_k(\mathbf{z}_k - \hat{\mathbf{z}}_{k|k-1}) \tag{7.30}$$

$$\mathbf{P}_{k|k} = \mathbf{P}_{k|k-1} - \mathbf{K}_k \mathbf{P}_{zz,k|k-1} \mathbf{K}_k^T \tag{7.31}$$

where the Kalman gain, $\mathbf{K}_k$, is

$$\mathbf{K}_k = \mathbf{P}_{xz,k|k-1} \mathbf{P}_{zz,k|k-1}^{-1}. \tag{7.32}$$

7.3 State-Dependent Riccati Equation (SDRE) Observers

A nonlinear system can be represented in a "state-dependent linear system" form Cloutier (1997). For such type of state-dependent systems one can adopt or extend the concepts of linear systems theory to nonlinear systems. The State-dependent Riccati Equation (SDRE) observer (Cloutier 1997; Cimen 2008; Shamma and Cloutier 2003) is a nonlinear observer which has been developed for the nonlinear systems. To apply such an estimation approach, first the nonlinear system is represented in the state-dependent coefficient (SDC) form and then the observer is synthesized for the state-dependent system. One of the main advantages of SDRE observer over other existing observers/estimators is that there exists infinitely many possible combinations of SDC systems (Cimen 2008; Lam et al. 2012). One can choose the best possible SDC system as per the problem at hand. Further, the SDRE observer does not need Jacobians during the estimation process.

Consider the discrete-time, process and measurement models

$$\mathbf{x}_k = \mathbf{f}(\mathbf{x}_{k-1}, \mathbf{u}_{k-1}) + \mathbf{w}_{k-1} \tag{7.33}$$
$$\mathbf{z}_k = \mathbf{h}(\mathbf{x}_k) + \mathbf{v}_k, \tag{7.34}$$

where, k is the time index showing the current time instant, $\mathbf{x}_k \in \mathbb{R}^n$ represents the state vector, $\mathbf{u}_k \in \mathbb{R}^q$ is the control input vector, $\mathbf{z}_k \in \mathbb{R}^p$ is the measurement vector, and $\mathbf{w}_{k-1}$ and $\mathbf{v}_k$ are the process and measurement noises. In this section, the process and measurement noises are assumed to be Gaussian with zero means and with covariances of $\mathbf{Q}_{k-1}$ and $\mathbf{R}_k$,

$$\mathbf{w}_{k-1} = \mathcal{N}(\mathbf{0}, \mathbf{Q}_{k-1})$$
$$\mathbf{v}_k = \mathcal{N}(\mathbf{0}, \mathbf{R}_k),$$

where, $\mathcal{N}$ represents the Gaussian or normal probability distribution.

The nonlinear models in Eqs. (7.33) and (7.34) can be represented in the state-dependent coefficient (SDC) (Mracek et al. 1996) form as

$$\mathbf{x}_k = \mathbf{F}(\mathbf{x}_{k-1})\mathbf{x}_{k-1} + \mathbf{G}(\mathbf{x}_{k-1})\mathbf{u}_{k-1} + \mathbf{w}_{k-1} \tag{7.35}$$
$$\mathbf{z}_k = \mathbf{H}(\mathbf{x}_k)\mathbf{x}_k + \mathbf{v}_k. \tag{7.36}$$

Note that the SDC matrices, **F**, **G** and **H**, for nonlinear systems are non-unique.

Consider a simple example, where the nonlinear plant dynamics is given by

$$\mathbf{x}_k = \mathbf{f}(\mathbf{x}_{k-1}) = \begin{bmatrix} x_{2,k-1} \\ x_{1,k-1}^2 \end{bmatrix}. \tag{7.37}$$

The SDC form of Eq. (7.37) can be written as:
$\mathbf{f}_{k-1} = \mathbf{F}_1(\mathbf{x}_{k-1})\mathbf{x}_{k-1} = \mathbf{F}_2(\mathbf{x}_{k-1})\mathbf{x}_{k-1} = \mathbf{F}_3(\mathbf{x}_{k-1})\mathbf{x}_{k-1} = \ldots$, where the SDC matrices are

$$\begin{aligned}
\mathbf{F}_1(\mathbf{x}_{k-1}) &= \begin{bmatrix} 0 & 1 \\ x_{1,k-1} & 0 \end{bmatrix}, \\
\mathbf{F}_2(\mathbf{x}_{k-1}) &= \begin{bmatrix} \frac{x_{2,k-1}}{x_{1,k-1}} & 0 \\ x_1 & 0 \end{bmatrix}, \\
\mathbf{F}_3(\mathbf{x}_{k-1}) &= \lambda \mathbf{F}_1(\mathbf{x}_{k-1})\mathbf{x}_{k-1} + (1-\lambda)\mathbf{F}_2(\mathbf{x}_{k-1}).
\end{aligned}$$

A similar example was considered in Banks et al. (2007). Different parameterisations can be obtained for different values of "λ" in the interval [0, 1], which satisfies the convexity of the chosen parametrisation. Similarly, different combinations of the SDC measurement matrices, $\mathbf{H}_i$, and the state vector, $\mathbf{x}_k$, can yield the same $\mathbf{h}(\mathbf{x}_k)$, like $\mathbf{H}_1(\mathbf{x}_k)\mathbf{x}_k = \mathbf{H}_2(\mathbf{x}_k)\mathbf{x}_k = \mathbf{H}_3(\mathbf{x}_k)\mathbf{x}_k = \mathbf{h}(\mathbf{x}_k)$.

As the SDC matrices are non-unique, before designing the SDRE observer (or SDRE controller), one should make sure that the state-dependent observability matrix (or the state-dependent controllability matrix) is of full rank (Banks et al. 2007).

Similar to the Kalman filter, the SDRE observer can be formulated in prediction and measurement update stages (Jaganath et al. 2005). The predicted state, $\hat{\mathbf{x}}_{k|k-1}$, and positive definite matrix, $\mathbf{P}_{k|k-1}$, can be written as:

$$\widehat{\mathbf{x}}_{k|k-1} = \mathbf{f}(\hat{\mathbf{x}}_{k-1|k-1}, \mathbf{u}_{k-1}) \tag{7.38}$$

$$\mathbf{P}_{k|k-1} = \mathbf{F}(\hat{\mathbf{x}}_{k-1|k-1})\mathbf{P}_{k-1|k-1}\mathbf{F}(\hat{\mathbf{x}}_{k-1|k-1})^T + \mathbf{Q}_{k-1}. \tag{7.39}$$

The updated state, $\hat{\mathbf{x}}_{k|k}$, and positive definite matrix, $\mathbf{P}_{k|k}$, are

$$\hat{\mathbf{x}}_{k|k} = \hat{\mathbf{x}}_{k|k-1} + \mathbf{K}_k[\mathbf{z}_k - \mathbf{H}(\hat{\mathbf{x}}_{k|k-1})] \tag{7.40}$$

$$\mathbf{P}_{k|k} = (\mathbf{I}_n - \mathbf{K}_k\mathbf{H}(\hat{\mathbf{x}}_{k|k-1}))\mathbf{P}_{k|k-1}, \tag{7.41}$$

where, $\mathbf{I}_n$ denotes the identity matrix of dimension $n \times n$ and the SDRE observer gain is

$$\mathbf{K}_k = \mathbf{P}_{k|k-1}\mathbf{H}(\hat{\mathbf{x}}_{k|k-1})^T \left[\mathbf{H}(\hat{\mathbf{x}}_{k|k-1})\mathbf{P}_{k|k-1}\mathbf{H}(\hat{\mathbf{x}}_{k|k-1})^T + \mathbf{R}_k\right]^{-1}. \tag{7.42}$$

It can be noted that the matrices $\mathbf{P}_{k|k-1}$ and $\mathbf{P}_{k|k}$ satisfy the Riccati recursion and are state-dependent. At first glance, the above formulation looks similar to that of the EKF; but actually it is not. The main difference between the SDRE observer and the EKF is the way in which the matrices $\mathbf{F}(\hat{\mathbf{x}}_{k-1|k-1}))$ and $\mathbf{H}(\hat{\mathbf{x}}_{k|k-1})$ are defined. In the SDRE observer they are the parameterised SDC matrices and for the EKF they are the Jacobians of the process and measurement models. The SDRE filter avoids the evaluation of Jacobians and hence avoids the errors introduced by the truncation of Taylor series. It is interesting to note that different parameterisations in the SDRE filter may yield different results (Liang and Lin 2013).

7.4 SDRE Information Filter

This section presents the formulation for the SDRE information filter (SDREIF). The SDRE filter and the extended information filter (EIF) are to be fused to form the SDREIF, which will have the benefits of both filters. The SDREIF is a derivative-free (Jacobian free) filter like the SDRE filter, and like the EIF, it is handy to deal with the multi-sensor state estimation. Similar to the information filter, the SDREIF is developed in the information domain, where the information vector and information matrix are propagated.

Consider the discrete-time process and measurement models given in Eqs. (7.33) and (7.34), and its SDC form in Eqs. (7.66) and (7.67). The predicted information vector, $\hat{\mathbf{y}}_{k|k-1}$, and the inverse of the positive definite matrix, $\mathbf{Y}_{k|k-1} = \mathbf{P}^{-1}_{k|k-1}$, for the SDREIF are:

$$\hat{\mathbf{y}}_{k|k-1} = \mathbf{Y}_{k|k-1}\hat{\mathbf{x}}_{k|k-1} \tag{7.43}$$

$$\mathbf{Y}_{k|k-1} = \left[\mathbf{F}(\hat{\mathbf{x}}_{k-1|k-1})\mathbf{Y}^{-1}_{k-1|k-1}\mathbf{F}(\hat{\mathbf{x}}_{k-1|k-1})^T + \mathbf{Q}_{k-1}\right]^{-1}, \tag{7.44}$$

where

$$\hat{\mathbf{x}}_{k|k-1} = \mathbf{f}(\hat{\mathbf{x}}_{k-1|k-1}, \mathbf{u}_{k-1}). \tag{7.45}$$

It can be seen that the predicted information vector in Eq. (7.43) is the same as that of the EIF, but the expression for $\mathbf{P}_{k|k-1}$ is different. In the EIF, the updated information matrix is the function of the Jacobian, $\nabla\mathbf{f}_x$, whereas in the SDREIF the $\nabla\mathbf{f}_x$ is replaced by the SDC $\mathbf{F}(\mathbf{x}_{k-1})$.

The updated information vector, $\hat{\mathbf{y}}_{k|k}$, and $\mathbf{Y}_{k|k}$ are:

$$\hat{\mathbf{y}}_{k|k} = \hat{\mathbf{y}}_{k|k-1} + \mathbf{i}_k \tag{7.46}$$

$$\mathbf{Y}_{k|k} = \mathbf{Y}_{k|k-1} + \mathbf{I}_k. \tag{7.47}$$

The information vector contribution, $\mathbf{i}_k$, and its associated information matrix contribution, $\mathbf{I}_k$, are

$$\mathbf{i}_k = \mathbf{H}(\hat{\mathbf{x}}_{k|k-1})^T \mathbf{R}_k^{-1} \left[\nu_k + \mathbf{H}(\hat{\mathbf{x}}_{k|k-1}) \hat{\mathbf{x}}_{k|k-1} \right] \tag{7.48}$$

$$\mathbf{I}_k = \mathbf{H}(\hat{\mathbf{x}}_{k|k-1})^T \mathbf{R}_k^{-1} \mathbf{H}(\hat{\mathbf{x}}_{k|k-1}) \tag{7.49}$$

where the measurement residual, ν_k, is

$$\nu_k = \mathbf{z}_k - \mathbf{H}(\hat{\mathbf{x}}_{k|k-1}). \tag{7.50}$$

7.4.1 SDREIF in Multi-sensor State Estimation

One of the main advantages of the state estimation in the information domain is its ability to easily deal with the multi-sensor state estimation (Mutambara 1998; Bar-Shalom et al. 2004). In the update stage, the information from various sensors can be easily fused (Mutambara 1998; Bar-Shalom et al. 2004). The prediction step for the SDREIF multi-sensor state estimation is the same as that of the SDREIF (with a single sensor). In the update stage, the data from different sensors are fused for an efficient and reliable estimation (Raol and Girija 2002).

Let the different sensors used for the state estimation be given by

$$\mathbf{z}_{j,k} = \mathbf{h}_{j,k}(\mathbf{x}_{j,k}) + \mathbf{v}_{j,k}; \qquad j = 1, 2, ...D, \tag{7.51}$$

where 'D' is the number of sensors. The parameterised SDC form for the multi-sensor models can be written as

$$\mathbf{z}_{j,k} = \mathbf{H}(\mathbf{x}_{j,k})\mathbf{x}_{j,k-1} + \mathbf{v}_{j,k}; \qquad j = 1, 2, ...D. \tag{7.52}$$

The updated information vector and the corresponding matrix for the multi-sensor SDREIF are

$$\hat{\mathbf{y}}_{k|k} = \hat{\mathbf{y}}_{k|k-1} + \sum_{j=1}^{D} \mathbf{i}_{j,k} \tag{7.53}$$

$$\mathbf{Y}_{k|k} = \mathbf{Y}_{k|k-1} + \sum_{j=1}^{D} \mathbf{I}_{j,k}, \tag{7.54}$$

where,

$$\mathbf{i}_{j,k} = \mathbf{H}(\hat{\mathbf{x}}_{j,k|k-1})^T \mathbf{R}_{j,k}^{-1} \left[\nu_k + \mathbf{H}(\hat{\mathbf{x}}_{j,k|k-1}) \hat{\mathbf{x}}_{j,k|k-1} \right] \tag{7.55}$$

$$\mathbf{I}_{j,k} = \mathbf{H}(\hat{\mathbf{x}}_{j,k|k-1})^T \mathbf{R}_{j,k}^{-1} \mathbf{H}(\hat{\mathbf{x}}_{j,k|k-1}) \tag{7.56}$$

The SDREIF, for both a single sensor and a set of multiple sensors, is summarised in Algorithm 7.3.

Algorithm 7.3 State-Dependent Riccati Equation Information Filter (SDREIF)

State-Dependent Coefficient Form:

Use paramerisation to bring the nonlinear process and measurement models in to the below SDC form:

$$\mathbf{x}_k = \mathbf{F}(\mathbf{x}_{k-1})\mathbf{x}_{k-1} + \mathbf{G}(\mathbf{x}_{k-1})\mathbf{u}_{k-1} + \mathbf{w}_{k-1}$$
$$\mathbf{z}_k = \mathbf{H}(\mathbf{x}_k)\mathbf{x}_{k-1} + \mathbf{v}_k.$$

Prediction:

Initialise the information vector, $\hat{\mathbf{y}}_{k-1|k-1}$, and the associated matrix, $\mathbf{Y}_{k-1|k-1}$; by setting $k = 1$.

The predicted information vector and the associate matrix are

$$\hat{\mathbf{y}}_{k|k-1} = \mathbf{Y}_{k|k-1}\hat{\mathbf{x}}_{k|k-1}$$
$$\mathbf{Y}_{k|k-1} = \left[\mathbf{F}(\hat{\mathbf{x}}_{k-1|k-1})\mathbf{Y}^{-1}_{k-1|k-1}\mathbf{F}(\hat{\mathbf{x}}_{k-1|k-1})^T + \mathbf{Q}_{k-1}\right]^{-1},$$

where,

$$\hat{\mathbf{x}}_{k|k-1} = \mathbf{f}(\hat{\mathbf{x}}_{k-1|k-1}, \mathbf{u}_{k-1}).$$

Measurement Update:

Evaluate the information vector contribution and its associated information matrix

$$\mathbf{i}_k = \mathbf{H}(\hat{\mathbf{x}}_{k|k-1})^T\mathbf{R}_k^{-1}\left[\nu_k + \mathbf{H}(\hat{\mathbf{x}}_{k|k-1})\hat{\mathbf{x}}_{k|k-1}\right] \tag{7.57}$$

$$\mathbf{I}_k = \mathbf{H}(\hat{\mathbf{x}}_{k|k-1})^T\mathbf{R}_k^{-1}\mathbf{H}(\hat{\mathbf{x}}_{k|k-1}) \tag{7.58}$$

where, the measurement residual, ν_k, is

$$\nu_k = \mathbf{z}_k - \mathbf{H}(\hat{\mathbf{x}}_{k|k-1}). \tag{7.59}$$

The updated information vector and the corresponding matrix are

$$\hat{\mathbf{y}}_{k|k} = \hat{\mathbf{y}}_{k|k-1} + \mathbf{i}_k$$
$$\mathbf{Y}_{k|k} = \mathbf{Y}_{k|k-1} + \mathbf{I}_k.$$

For the state estimation with multiple sensors, the updated information vector and the corresponding matrix are

$$\hat{\mathbf{y}}_{k|k} = \hat{\mathbf{y}}_{k|k-1} + \sum_{j=1}^{D}\mathbf{i}_{j,k} \tag{7.60}$$

$$\mathbf{Y}_{k|k} = \mathbf{Y}_{k|k-1} + \sum_{j=1}^{D}\mathbf{I}_{j,k}, \tag{7.61}$$

where,

$$\mathbf{i}_{j,k} = \mathbf{H}(\hat{\mathbf{x}}_{j,k|k-1})^T\mathbf{R}_{j,k}^{-1}\left[\nu_k + \mathbf{H}(\hat{\mathbf{x}}_{j,k|k-1})\hat{\mathbf{x}}_{j,k|k-1}\right] \tag{7.62}$$

$$\mathbf{I}_{j,k} = \mathbf{H}(\hat{\mathbf{x}}_{j,k|k-1})^T\mathbf{R}_{j,k}^{-1}\mathbf{H}(\hat{\mathbf{x}}_{j,k|k-1}) \tag{7.63}$$

Recovery of State and the Corresponding Matrix:

The state and the corresponding positive definite matrix can be recovered as

$$\hat{\mathbf{x}}_{k|k} = \mathbf{Y}_{k|k} \backslash \hat{\mathbf{y}}_{k|k} \tag{7.64}$$

$$\mathbf{P}_{k|k} = \mathbf{Y}_{k|k} \backslash \mathbf{I}_n \tag{7.65}$$

where $\mathbf{I}_n$ is the state vector sized identity matrix.

7.5 PMSM Case Revisited

The state estimation of a two phase permanent magnet synchronous motor (PMSM) was considered in Chap. 6 and is to be reconsidered here by using the state-dependent Riccati equation information filtering approach. The state vector of the PMSM are, $[i_a, i_b, \omega, \theta]^T$. The inputs to the PMSM are the voltages, $u_{1,k}$ and $u_{2,k}$. It is assumed that the first two states, i_a and i_b, are available for direct measurement and the remaining two states, ω and θ, are to be estimated.

The discrete-time nonlinear model of PMSM is Simon (2006), Chandra et al. (2013)

$$\begin{bmatrix} i_{a,k+1} \\ i_{b,k+1} \\ \omega_{k+1} \\ \theta_{k+1} \end{bmatrix} = \begin{bmatrix} i_{a,k} + T_s(-\frac{R}{L} i_{a,k} + \frac{\omega\lambda}{L} \sin\theta_k + \frac{1}{L} u_{1,k}) \\ i_{b,k} + T_s(-\frac{R}{L} i_{b,k} - \frac{\omega\lambda}{L} \cos\theta_k + \frac{1}{L} u_{2,k}) \\ \theta_k + T_s(-\frac{3\lambda}{2J} i_{a,k} \sin\theta_k + \frac{3\lambda}{2J} i_{b,k} \cos\theta_k - \frac{F\omega_k}{J}) \\ \theta_k + T_s\omega_k \end{bmatrix}$$

the outputs and inputs are

$$\begin{bmatrix} y_{1,k} \\ y_{2,k} \end{bmatrix} = \begin{bmatrix} i_{a,k} \\ i_{b,k} \end{bmatrix}, \quad \begin{bmatrix} u_{1,k} \\ u_{2,k} \end{bmatrix} = \begin{bmatrix} \sin(0.002\pi k) \\ \cos(0.002\pi k) \end{bmatrix}.$$

The following parameters are used for the simulations (Simon 2006; Chandra et al. 2013) R = 1.9 Ω, λ = 0.1, L = 0.003 H, J = 0.00018, F = 0.001 and $T_s = 0.001$ s. It is assumed that the plant and measurement models are excited with the additive Gaussian noises.

The first step to implement the SDREIF is to bring the process and measurement equations in the SDC form. In this section the following SDC form is to be used:

$$\mathbf{x}_k = \mathbf{F}(\mathbf{x}_{k-1})\mathbf{x}_{k-1} + \mathbf{G}(\mathbf{x}_{k-1})\mathbf{u}_{k-1} + \mathbf{w}_{k-1} \tag{7.66}$$

$$\mathbf{z}_k = \mathbf{H}(\mathbf{x}_k)\mathbf{x}_{k-1} + \mathbf{v}_k. \tag{7.67}$$

where,

$$\mathbf{F}(\mathbf{x}_{k-1}) = \mathbf{I}_4 + T_s \begin{bmatrix} -\frac{R}{L} & 0 & \frac{\lambda}{L}\sin\theta_k & 0 \\ 0 & -\frac{R}{L} & \frac{\lambda}{L}\cos\theta_k & 0 \\ -\frac{3}{2}\frac{\lambda}{J}\sin\theta_k & \frac{3}{2}\frac{\lambda}{J}\cos\theta_k & -\frac{B}{J} & 0 \\ 0 & 0 & 1 & 0 \end{bmatrix}$$

$$\mathbf{G}(\mathbf{x}_{k-1}) = \begin{bmatrix} \frac{1}{L} & 0 \\ 0 & \frac{1}{L} \\ 0 & 0 \\ 0 & 0 \end{bmatrix}$$

$$\mathbf{H}(\mathbf{x}_k) = \begin{bmatrix} 1 & 0 & 0 & 0 \\ 0 & 1 & 0 & 0 \end{bmatrix}$$

Two groups of simulations have been performed on the PMSM to analyse the proposed SDREIF. The first group is simulated with a single set of sensors and the second group of simulations are performed with two sets of sensors. In the first group of simulations, the covariance of the process and measurement noise is

$$\mathbf{Q} = \begin{bmatrix} 11.1111 & 0 & 0 & 0 \\ 0 & 11.1111 & 0 & 0 \\ 0 & 0 & 0.0025 & 0 \\ 0 & 0 & 0 & 1\times 10^{-6} \end{bmatrix}, \quad \mathbf{R} = 1\times 10^{-4}\mathbf{I}_2$$

The initial information vector is selected from $\mathcal{N}\left(\begin{bmatrix}1 & 1 & 1 & 1\end{bmatrix}^T, \mathbf{I}_4\right)$.

It was further assumed that the first sensor (i_a measurement) and the second sensor (i_b measurement) are faulty from 0.5–0.8 s and 2–2.3 s, respectively. In this case study, the term 'faulty' means the sensor only outputs zero measurements during this condition. Figure 7.2 shows the plots of actual and measured data; where it was assumed that the first and second sensors are faulty for some time periods. In the first group of simulations, a single set of sensors are used to measure the currents, i_a and i_b; and ω and θ are estimated using the SDREIF given in Sect. 7.4 and for the second group of simulations, two sets of sensors are used and the procedure given in the Sect. 7.4.1 is used. For the multi-sensor state estimation, the first set of sensors are the same as that of the single-sensor state estimation; the noise covariance of the second set of sensors is: $\mathbf{R}_2 = 4\times 10^{-6}\mathbf{I}_2$. The corresponding results are depicted in Fig. 7.3. It can be seen that the SDREIF using a single set of sensors are not capable to handle the faulty sensors; whereas the multi-sensor SDREIF can handle it efficiently. It is interesting to note that, during the faulty measurements also the multi-sensor SDREIF could able to keep the error within an acceptable bound. The error plots using the single-set and multi-set of sensors are shown in Fig. 7.4.

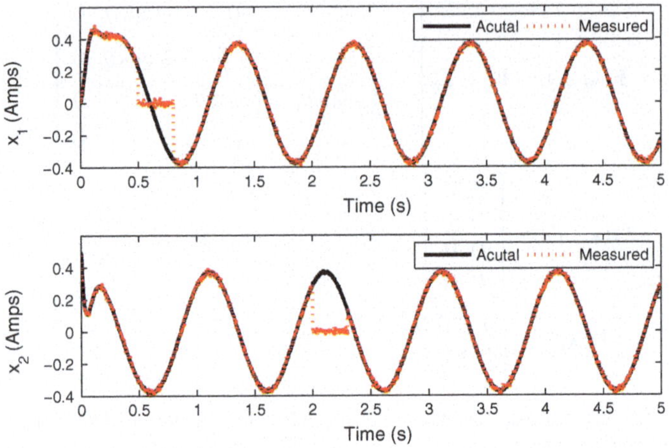

Fig. 7.2 Actual and measured data

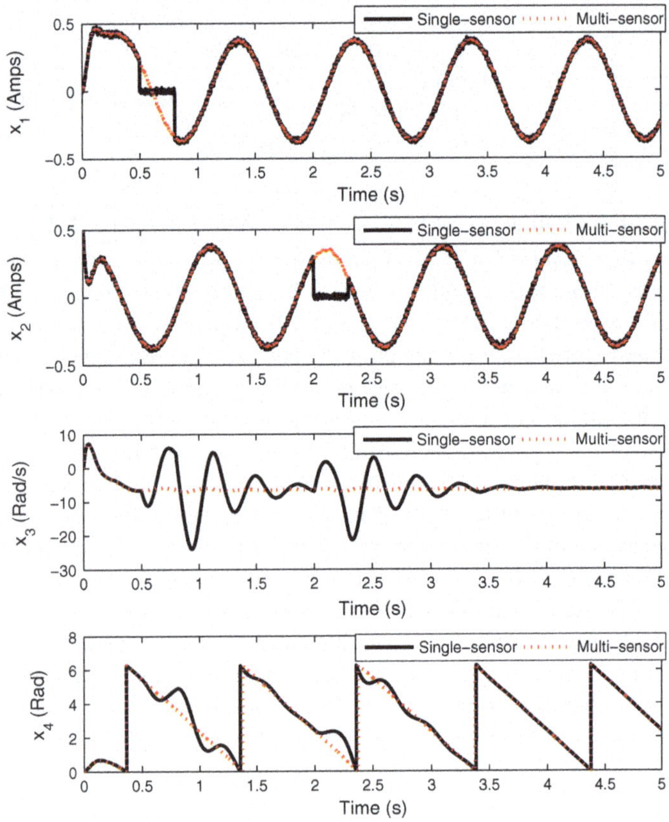

Fig. 7.3 State estimation using single- and multi-sensor SDREIF

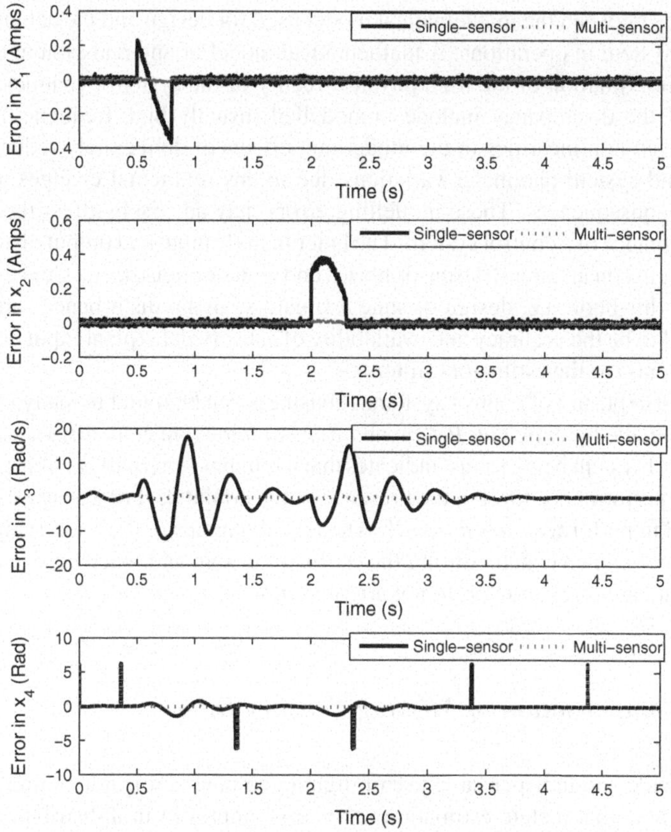

Fig. 7.4 Error plots using single- and multi-sensor SDREIF

7.6 Filter Robustness Consideration

All engineering control systems are operating in the real environment, and are subject to uncertainties. Therefore, a practically useful control system must have the ability to deal with uncertainties. This is so-called "robustness" property. This section will first introduce what kind of uncertainties commonly affect the behaviour of a control system, and then will consider the robustness of state estimators (filters).

7.6.1 Uncertainties and Robustness Requirement

It is well understood that uncertainties are unavoidable for a real control system in its operation. The uncertainty possibly affecting a control system can be classified in two categories: dynamic perturbations and signal disturbances. The former represent the discrepancy between the mathematical model used for design and the actual dynamics of the system in operations. A mathematical model of any real system is always just an approximation of the true, physical reality of the system dynamics. Typical sources of the discrepancy include unmodelled (usually high-frequency) dynamics, neglected nonlinearities in the modelling, effects of deliberately reduced-order models, and system-parameter variations due to environmental changes and torn-and-worn consequences. These modelling errors may adversely affect the stability and performance of a control system. The latter include input and output disturbances (such as a gust on an aircraft), sensor noises and actuator noises, etc. Concerning the subject of this book, i.e. design of state estimators, signal disturbances are mainly considered to be the accuracy and availability of sensory data (plant input and output measurements) to the estimators (filters).

Detailed exposure of control systems robustness can be found in many resources, for example, the book by *Gu, Petkov and Konstantinov* (Gu et al. 2014).

A control system being robust indicates that it remains (internally) stable and keeps up certain performance level in the presence of uncertainties. For a state estimator, it is required to perform *nearly normally* when the dynamics of the underlying control system is not known exactly during the estimator design and when occasionally the sensory data are inaccurate or do not arrive in time.

7.6.2 Compensation of Missing Sensory Data

With the wide and indispensable use of digital computers, it is not an unusual phenomenon now that a state estimation scheme is employed in a distributed system structure. It is common that the plant, controller and filter are physically located at different places and communications (data transport) over a wireless network become necessary. Practical examples can be found in a ground-controlled unmanned air vehicle operation or in an oil field operation. In such cases, data loss or delay would be quite possible due to the limited spectrum of communication channels, time-varying channel gains, interference, congestions, limited bandwidth of buffer registers, transmission errors, etc. This subsection will introduce a few approaches to deal with such scenarios. It shall mainly discuss the case of plant (output) measurement data loss as this is the most common scenario. Also, for the simplicity of exposure, the conventional Kalman filter (KF) algorithm shall be used in the discussion. The approaches discussed next can be however applied in plant input data loss scenario and/or with other filtering algorithms, either linear or nonlinear, with appropriate modifications.

There are various approaches to compensate the measurement signal in the case of data loss. For example, see Khan (2012), Khan et al. (2010), Schenato (2008), Liu and Goldsmith (2004), Sinopoli et al. (2004), Shi et al. (2005) for details and analysis. The aim in this section is to find appropriate values for a signal during the future time period, though it could be short, when the signal is not available. This is a prediction of signal values. Due to the approaches described below being mainly in linear equations domain, it is called *linear prediction* (LP) (Vaidyanathan 2008). The technique and terminology are widely used in control system identification processes where signals are reconstructed from its previous dynamic signal samples (Makhoul 1975). In the mathematics domain, linear prediction is a mathematical operation where future values of a discrete-time signal are estimated as a linear function of existing (available) samples and it is in the sense of “extrapolation”.

Calculation of linear prediction coefficients (LPC) is based on the following assumptions:

- The signal has correlation between its samples.
- The statistical properties of the signal vary slowly with time.

If the above assumptions are satisfied, the future values, or the current value depending on the definition of the index k, of the signal can be reconstructed (Chu 2004). In the context considered, the scheme can be represented by the following equation:

$$\bar{z}_k = \sum_{j=1}^{n} \alpha_j z_{k-j} + \sum_{i=0}^{N} b_i u_{k-i} \tag{7.68}$$

where $\bar{z}_k$ denotes the estimated value at the time instant k; and, $\alpha_j = [\alpha_1, \alpha_2, \ldots.\alpha_n]$ are parameters of the algorithm which are so-called Linear Prediction parameters of measurement (output) and $b_i = [b_1, b_2, \ldots.b_n]$ are the parameters of the input. Based on various combinations, one might define different prediction schemes in relation to the linear prediction method, e.g. Autoregressive (AR) Model, Moving Average (MA) Model, Autoregressive Moving Average (ARMA) Model etc., see (Vaidyanathan 2007) for further details. Obviously, the value of linear prediction coefficients (LPC) and the orders of the two terms in the right side of (7.68) affects the result of compensation greatly. That will be discussed a bit further later in this chapter.

Zero Kalman Gain Matrix

When the measurement is not available, a straightforward idea is just to ignore it and to calculate the “measurement” directly from the system output equation with newly available predicted state instead of true state. That is, in the Kalman filter algorithm, one assumes that $\mathbf{z}_k = \mathbf{H}_k \widehat{\mathbf{x}}_{k|k-1}$. That leads to the residual (innovation) vector ν_k to be zero, i.e. the “measurement” equals the predicted output. It is equivalent to setting the Kalman filter gain matrix $\mathbf{K}_k$ to be zero consequently in the updating stage. As soon as the measurement becomes available, the calculations go back to the normal procedure in the Kalman filter algorithm.

Despite of fast calculation, simpler structure and no need of storing observations, there are quite a few drawbacks associated with this approach for the loss of observation. In practice, Kalman filter procedure with zero gain matrix may diverge and it is likely to happen that the error covariance could exceed the bounded error limits provided (Liu and Goldsmith 2004). Another shortcoming is that when the system measurement (observation) is resumed after the loss period, one can see a sharp spike phenomenon in the estimation error, a kind of bump switch in control engineering terminology. This is because the Kalman filter gain is set to zero when data loss is occurred; however, when observation is repossessed, the filter gain will surge to a high value and then attain the steady state value in order to compensate loss impact (Khan and Gu 2009). This feature consequently results in a sudden peak to reach the normal trajectory of the estimated state which is an undesirable behaviour of the estimation algorithm. Another serious shortcoming associated with zeroing Kalman gain matrix approach is that it takes a far long time to regain the values of estimated states and covariance within satisfactory bands afterwards. In theory, it will take infinite time to retrieve the steady state values, due to the recursive nature of the filtering algorithm operations.

Zero-order Hold (ZOH)

If the filter parameters are selected as $n = 1$, $\alpha = 1$ and without the input terms in Eq. (7.68), the compensated observation vector will be retranslated as

$$\bar{z}_k = z_{k-1} = Hx_{k-1} + v_{k-1}$$

In the literature, the above scheme is known as the Zero-order Hold (ZOH) method, due to the fact that the last observation is held to provide the measurement update step for the predicted state and error covariance. The ZOH method can be a useful approach for systems with sufficiently slow measurement samples (such as ship dynamics, commercial airplanes, etc.). However, it is prone to fail in the reconstruction of the given signal if the sampling frequency of sensors measurements are high (such as with jet fighters, UAVs, etc.). It also will not work when the data loss continues to a relatively long period of time.

First-order Hold (FOH)

If it is assumed that $n = 2$ and $\alpha_i = [\alpha_1, \alpha_2]$, and without the input terms in Eq. (7.68), the following will be obtained

$$\bar{z}_k = \alpha_1 z_{k-1} + \alpha_2 z_{k-2} \tag{7.69}$$

The scheme is known as the First-order Hold (FOH) approach. When $\alpha_1 = \alpha_2 = 0.5$, i.e. the two weights are set equal, it will become

$$\bar{z}_k = \frac{z_{k-1} + z_{k-2}}{2}$$

As it can be seen, in the above LP approaches, one does not require to compute the parameters in relation to the linear prediction system. Obviously, these two coefficients can be set differently to form weighted observations. That is to view the previous two measurements, (z_{k-1} and z_{k-2}), with different significance on forming the current, estimated measurement. Usually and logically, the closer data would have more impact on the estimated measurement and hence the bigger weight to be set.

Trend Approach

In some applications, one may notice that the observation vector contains a regular pattern in the received data. In such cases, data missing would not affect the estimation to a high extent while these trends can vary from time to time and from a system to another (Kanjilal 1995). For the sake of easy understanding, and without loss of generality, it is assumed that

$$\bar{z}_k = z_{k-1} + [z_{k-1} - z_{k-2}] \tag{7.70}$$

or

$$\begin{aligned}
\bar{z}_{k+1} &= Hx_k + v_k + [Hx_k + v_k - Hx_{k-1} - v_{k-1}] \\
&= Hx_k + v_k + [H\Delta_{xk} + \Delta_{vk}] \\
&= H(x_k + \Delta_{xk}) + (v_k + \Delta_{vk}) \\
&= H\bar{x}_{k+1} + \bar{v}_{k+1}
\end{aligned} \tag{7.71}$$

where $\Delta_{xk} = x_k - x_{k-1}$ and $\Delta_{vk} = v_k - v_{k-1}$, $\bar{x}_{k+1} = x_k + \Delta_{xk}$ and $\bar{v}_{k+1} = v_k + \Delta_{vk}$.

Autoregressive Approach (AR)

For $n \geq 2$ in (7.68) and without the input terms, the approach is termed as Autoregressive (AR) approach in which every (past) measurement is associated with a weight (coefficient) to form the current measurement(estimation). All the three approaches above are actually special cases of the Autoregressive approach. Therefore, the measurement equation is written as

$$\begin{aligned}
\bar{z}_k &= \frac{1}{n}\sum_{j=1}^{n} z_{k-j} \\
&= H\frac{1}{n}\sum_{j=1}^{n} x_{k-j} + \frac{1}{n}\sum_{j=1}^{n} v_{k-j} \\
&= H\bar{x}_k + \bar{v}_k
\end{aligned} \tag{7.72}$$

where $\bar{x}_k = \frac{1}{n}\sum_{j=1}^{n} x_{k-j}$ and $\bar{v}_k = \frac{1}{n}\sum_{j=1}^{n} v_{k-j}$.

Autoregressive Moving Average Approach (ARMA)

The approaches introduced above are all concentrating on the past sensed, or past estimated, measurements. They do not depend on the input signal to the underlying system, i.e. the second part of the right-hand side of Eq. (7.68) is not explored. The autoregressive moving average (ARMA) approach however does explore the use of the input. With the ARMA approach, more design freedom, in terms of number of design parameters, is available and may produce better estimation results.

All the design parameters involved in (7.68) can be determined by using some optimisation methods, for example the ones described in Khan et al. (2010). However, it should be noted that higher order (n and N in (7.68)) does not always imply better estimation results (Chu 2004).

7.6.3 Selection of Linear Prediction Coefficients and Orders

In this subsection a method to select the linear prediction coefficients and order of an autoregressive model used in observation data loss compensation is to be introduced. This method and the case study followed in the next subsection were firstly proposed and considered in Khan et al. (2010, 2012). Interested readers can find more details in these references. However, please note some notations and definitions of nomenclature have been changed to suit those used in this book.

It is assumed that a general AR-series is mathematically modelled by

$$\bar{z}_k = c + \sum_{j=1}^{n} \alpha_j z_{k-j} + \xi_k \tag{7.73}$$

where $\bar{z}_k$ is the signal which is to be reconstructed based on previous sensed data and hence it is referred as compensated output signal. To carry out the signal prediction, "optimal" values of α_j are to be computed through the auto-correlation function using Yule-Walker equation (Chu 2004):

$$\gamma_m = \sum_{j=1}^{n} \alpha_j \gamma_{m-j} + \sigma^2 \delta_m \tag{7.74}$$

where δ_m is the Kronecker delta, that is

$$\delta_m = \begin{cases} 1; \text{ if m = 0} \\ 0; \text{ otherwise} \end{cases} \tag{7.75}$$

In Eq. (7.73) c is a constant (normally the mean of the signal) and ξ_k is a noise input signal. If the input is a white noise, the model is referred to as an "All-Pole Infinite Impulse Response (IIR)" filter, where n is the filter order. It is assumed in this book that the noise involved is white and unmeasurable, and consequently c and the noise term will vanish in the equation. Hence it is supposed

$$\bar{z}_k = \sum_{j=1}^{n} \alpha_j z_{k-j} \tag{7.76}$$

The computation of "optimal" values of α_j and the order of the linear prediction (LP) filter, n, are critical for the estimation problem with loss of measurements. It is worthwhile to note that a large number of previous data does not always guarantee an optimal prediction of the missing data (Chu 2004).

In general, there are two types of processes: *First*- Stationary Process where the joint probability distribution of the process does not change with time shifting; *Secondly*- Non-stationary Process, i.e. it does change with time. In this book, only the stationary processes are considered. In addition, it is assumed that the process is Wide-Sense Stationary (WSS), i.e. the mean and variance are time independent (Hayes 2009), which is shown as

$$\begin{aligned} E\big[z(t)\big] &= m_z(t) = m_z(t+\tau)\ \forall\ \tau \in \mathbb{R} \\ E\big[z(t_1)z(t_2)\big] &= E\big[z(t_1+\tau)z(t_2+\tau)\big] \\ &= R_z(t_1, t_2)\ \ \forall\ \ \tau \in \mathbb{R} \end{aligned} \tag{7.77}$$

The linear prediction parameters, the coefficients and order, define the contributions of previous observation made for reconstructing the missing data. This can be viewed from the data autocorrelation standpoint. The output error generated by this compensated observation vector is

$$e_z(k) = \bar{z}_k - z_k \tag{7.78}$$

The residual related cost function can be defined based on the output error in the form of

$$J_k = E\big[e_z(k)^T e_z(k)\big] \tag{7.79}$$

Optimal values of the LP parameters, α_j can be found by minimizing the above cost function, J_k. That is,

$$\begin{aligned} \frac{\partial J_k}{\partial \alpha_i} &= \frac{\partial}{\partial \alpha_j} E\big[e_z(k)^T e_z(k)\big] \\ &= E\Big[\frac{\partial\big(e_z(k)^T e_z(k)\big)}{\partial \alpha_i}\Big] \\ &= E\Big[\frac{\partial J_k}{\partial \bar{z}_k}\frac{\partial \bar{z}_k}{\partial \alpha_i}\Big] \end{aligned} \tag{7.80}$$

Also, from Eqs. (7.78) and (7.79)

$$\frac{\partial J_k}{\partial \bar{z}_k} = 2\big[e_z(k)^T\big] = 2\big[\bar{z}_k - z_k\big]^T \tag{7.81}$$

By differentiating $\bar{z}_k$ with regard to α_i, one can have

$$\frac{\partial \bar{z}_k}{\partial \alpha_i} = z_{k-i} \tag{7.82}$$

In order to minimize the cost function, J_k, from Eq. (7.80), one should first have

$$\frac{\partial J_k}{\partial \alpha_i} = 0 \tag{7.83}$$

From Eqs. (7.80)–(7.82),

$$E\left[2\left(\bar{z}_k - z_k\right)^T\right]z_{k-i} = 0$$

$$E\left[\bar{z}_k^T z_{k-i}\right] - E\left[z_k^T z_{k-i}\right] = 0$$

By substituting Eq. (7.76), one will have

$$\sum_{j=1}^{n} \alpha_j E\left[z_{k-j}^T z_{k-i}\right] = E\left[z_k^T z_{k-i}\right] \tag{7.84}$$

which can be simplified as

$$R_\gamma A_\alpha = r_\gamma \tag{7.85}$$

or

$$A_\alpha = R_\gamma^{-1} \cdot r_\gamma \tag{7.86}$$

where

$$R_\gamma = \begin{bmatrix} R_\gamma(0) & R_\gamma(1) & R_\gamma(2) & \cdots & R_\gamma(n-1) \\ R_\gamma(1) & R_\gamma(0) & R_\gamma(1) & \cdots & R_\gamma(n-2) \\ R_\gamma(2) & R_\gamma(1) & R_\gamma(0) & \cdots & R_\gamma(n-3) \\ \vdots & \vdots & \vdots & \ddots & \vdots \\ R_\gamma(n-1) & R_\gamma(n-2) & R_\gamma(n-3) & \cdots & R_\gamma(0) \end{bmatrix} \tag{7.87}$$

$$A_\alpha = \left[\alpha_1\ \alpha_2\ \alpha_3\ \cdots\ \alpha_n\right]^T \tag{7.88}$$

$$r_\gamma = \left[r_\gamma(1)\ r_\gamma(2)\ r_\gamma(3)\ \cdots\ r_\gamma(n)\right]^T \tag{7.89}$$

$$E\left[z_{k-j}^T z_{k-i}\right] = \begin{cases} R_\gamma(0), & \text{if j = i} \\ R_\gamma(|j-i|), & \text{if j} \neq \text{i} \end{cases} \tag{7.90}$$

and

$$r_\gamma(i) = E\left[z_k^T z_{k-i}\right] \tag{7.91}$$

The LP filter optimal values will be calculated through solving Eq. (7.86). It is worthwhile to notice that the complexity of the LP filter is directly related to the calculation burdensome of Eq. (7.86) which depends heavily on the size of matrix R_γ, or the order of LP filter. Higher orders of an LP filter does not necessarily yield a better reconstruction of a lost signal as mentioned earlier. It is therefore useful to address the "optimal order" of the LP filter next.

As being discussed earlier, the higher orders of LP filter does not guarantee a better reconstruction of signal, which makes the selection process of the order of an LP filter more critical. Toward this end, it is essential to propose some easy-to-implement approaches in order to provide "good" values of the LP filter order. Among a few alternative algorithms, the following straightforward approaches are found "better" than other methods along with an assumption required which is listed below.

Assumption: To employ these methods, it is assumed that $T_f \geq nT_s$ where T_s is the sampling time, T_f is the starting time of loss of observations (LOOB), and n is the LP filter's order. This assumption is clearly required due to the fact that the lost observation requires n signal sample to be reconstructed as of Eq. (7.86). This assumption indicates an additional importance for the selection of appropriate LP order.

Assume that the starting time instant of LOOB to be at time $t = k$. Having the measurement updated step known at time instant $k - 1$, two methods are to be proposed both of which are aimed at yielding an optimal order of the LP filter. The first method is detailed in Algorithm 7.4.

Algorithm 7.4 Selection of the LP filter order (first method)

Compute $e_m(k-1) = \max(e_i)$, where $e_i = x_i - \hat{x}_i$ for
$i = \{1, 2, ...k - 1\}$.
Initialization $j = 1$, Compute R_γ and r_γ through Eqs. (7.90) and (7.91) respectively.
Recursion: $j = 2....M$
Obtain $\bar{z}$ through Eq. (7.76)
Calculate measurement updated state ${}_c\hat{x}_k$
based on this Compensated Observations
Compute $e_j(k) = x_k - \hat{x}_k$.
Check Is $e_j(k) \leq e_m(k-1)$
Yes $n \longleftarrow j$: Order of the LP Filter
Else $j \longleftarrow j + 1$
Repeat Step 4

It is worthwhile to emphasise that the autocorrelation of a function indicates in fact the similarity of the signal with its timely separated function. Hence, for a well-established R_γ, the above factors of R_γ are ranging from $+1$, for a perfect correlation, to -1, for a fully anti-correlated case. Hence the maximum correlation would be for zero time shift.

Generally speaking, the autocorrelation function for two different time shifts p and q will be

$$\gamma(p,q) = \frac{E\big[(z_{k-p} - \mu_p)(z_{k-q} - \mu_q)\big]}{\sigma_p \sigma_q} \tag{7.92}$$

Due to the property of WSS (7.77), for the same time shifts ($p = q$), the autocorrelation would result in

$$\begin{aligned}\gamma(p,p) = R_\gamma(p) &= \frac{E\big[(z_{k-p} - \mu)(z_{k-p} - \mu)\big]}{\sigma_{k-p}\sigma_{k-p}} \\ &= \frac{E\big[(z_{k-p} - \mu)(z_{k-p} - \mu)\big]}{\sigma^2} \\ &= 1\end{aligned} \tag{7.93}$$

for any non-zero variance σ value. It is also noticed that the longer the time shift is, the lesser correlation will be resulted. Hence, the following autocorrelation-related inequality can be readily inferred

$$R_\gamma(n) \geq \cdots R_\gamma(j) \geq \cdots \geq R_\gamma(1) \geq R_\gamma(0) \tag{7.94}$$

The inequality criteria of Eq. (7.94), most likely, leads to a simplest routine in choosing the order of LP filter. Nevertheless, one might propose alternative approaches to tackle the problem of order selection process. To this end, another algorithm is to be introduced which is based on the above inequality to extract the LP order and is summarized in Algorithm 7.5.

Algorithm 7.5 Selection of the LP filter order (second method)

Select γ_{th}.
Recursion $j = 1, 2, \cdots M$
Compute $\gamma(j)$ using Eq. (7.92)
Check: Is $\gamma(j) \leq \gamma_{th}$,
  Yes Stop further computation of γ_j and
    Select order of LP filter ($n \leftarrow j$)
  Else update $j \longleftarrow j + 1$
Repeat Step 3

7.6.4 A Case Study

In this section, the Mass-Spring Damper (MSD) system shown in Fig. 7.5 is to be modelled first. This is a slightly modified version of the MSD system analysed in Fekri et al. (2007). Four states are related with this system, out of which only one is measured through a noisy sensor. Initially the conventional Kalman filter is implemented and tested with no loss of data (normal operation). Thereafter, an (artificial) fault is introduced in the sensor due to which the sensor readings are not avail-

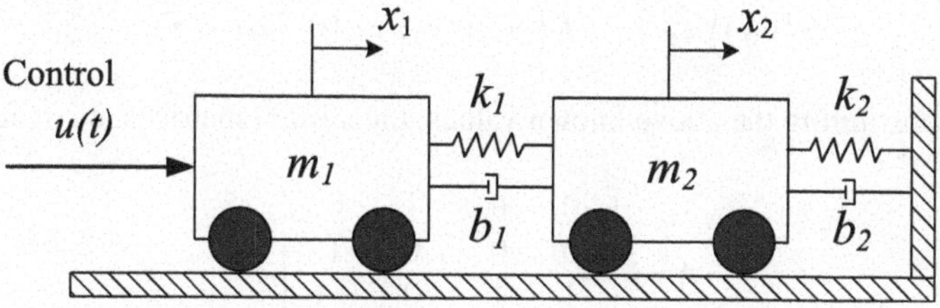

Fig. 7.5 MSD two cart system

able for the duration of 2.3–3.3 s. The compensation with zero Kalman gain matrix and then the AR compensation algorithm with optimal linear prediction coefficients introduced earlier will be applied in such scenarios to see the data compensation results.

Model of Mass-Spring-Damper (MSD) System

The system under study here is a mass-spring-damper (MSD) system, whose continuous-time dynamics are described as follows:

$$\begin{aligned} \dot{x}(t) &= Ax(t) + Bu(t) + Lw(t) \\ z_k &= Cx_k + v_k \end{aligned} \tag{7.95}$$

The state vector

$$x^T(t) = [\, x_1(t)\ x_2(t)\ \dot{x}_1(t)\ \dot{x}_2(t) \,] \tag{7.96}$$

is composed of displacements and speed of the two masses in Fig 7.5.

$$A = \begin{bmatrix} 0 & 0 & 1 & 0 \\ 0 & 0 & 0 & 1 \\ \frac{k_1}{m_1} & \frac{k_1}{m_1} & -\frac{b1}{m1} & \frac{b1}{m1} \\ \frac{k_1}{m_2} & -\frac{k_1+k_2}{m_1} & \frac{b_1}{m_2} & -\frac{b_1+b_2}{m_2} \end{bmatrix} \tag{7.97}$$

$$B^T = [0\,0\,\tfrac{1}{m_1}\,0] \tag{7.98}$$

$$C = [0\,1\,0\,0] \tag{7.99}$$

$$L = [0\,0\,0\,3] \tag{7.100}$$

The known parameters are $m_1 = m_2 = 1$, $k_1 = 1$, $k_2 = 0.15$, $b_1 = b_2 = 0.1$ and the sampling time is $T_s = 1$ ms. Also, the MSD plant disturbance and sensor noise dynamics are characterized as

$$E\{w(t)\} = 0, \quad E\{w(t)w(\tau)\} = \Xi\delta(t-\tau), \quad \Xi = 1 \tag{7.101}$$

$$E\{v(t)\} = 0, \quad E\{v(t)v(\tau)\} = 10^{-6}\delta(t-\tau) \tag{7.102}$$

After substituting the above known values, the system matrices are as follows:

$$A = \begin{bmatrix} 0 & 0 & 1 & 0 \\ 0 & 0 & 0 & 1 \\ 1 & 1 & -0.1 & 0.1 \\ 0.1 & -1.15 & 0.1 & -0.2 \end{bmatrix} \tag{7.103}$$

and

$$B^T = [0\,0\,1\,0] \tag{7.104}$$

The simulation results of using the proposed AR model data compensation algorithm to the above MSD system in the case of measurement data loss will be shown next.

For the purpose of the study, the continuous-time dynamics of the MSD system is transformed appropriately to a discrete-time model using the sampling time of $T_s = 0.001$ s. In the simulation results, all the four states are presented but the measured state (the second state) is specially highlighted to show the performance in the case with compensation and the case without compensation (i.e. no update in Kalman filtering with Kalman gain to be zero) during the loss of observation (LOOB) period. Also, the conventional Kalman filter (KF) performance is included, too. In the case of data loss, KF simply takes the measurement as zero value.

State Estimation

For the first set of simulations of the MSD case-study, the performance of the conventional Kalman filter (KF) is considered. Towards this end, the simulations are to be carried out by assuming that there is no measurement loss (i.e. the normal operation). The results of the normal operation case are shown in Figs. 7.6 and 7.7. Figure 7.6 shows the single noisy output measurement ($z(t)$, the position of mass M_2) along with its actual state, $x_2(t)$, and the associated state estimation, $\hat{x}_2(t)$. As expected in the failure-free case (the normal operation), the conventional discrete-time Kalman filter is able to successfully estimate the position of the mass M_2 within the provided simulation range. In Fig. 7.7 the three remaining states (namely, x_1, the position of mass M_1; x_3, the velocity of mass M_1; and x_4, the velocity of mass M_2) are illustrated together with their associated state estimations. It can be seen from Fig. 7.7 that, in the normal operational mode, KF is working perfectly, as all other states (which were not observed at the output of the system) are estimated reasonably well. It is worthwhile to notice that, as expected, the effect of the plant disturbance is more severe on the velocity of mass M_2 than on other states - this can be seen by some sort of juggling in the results of the fourth state, x_4, as illustrated in Fig. 7.7.

Now, a typical set of simulations subject to the fault-induced measurements will be considered. It is to compare the results of the compensated case using "optimal" LP parameters (ComKF) with those of the KF filter using zero Kalman gain matrix

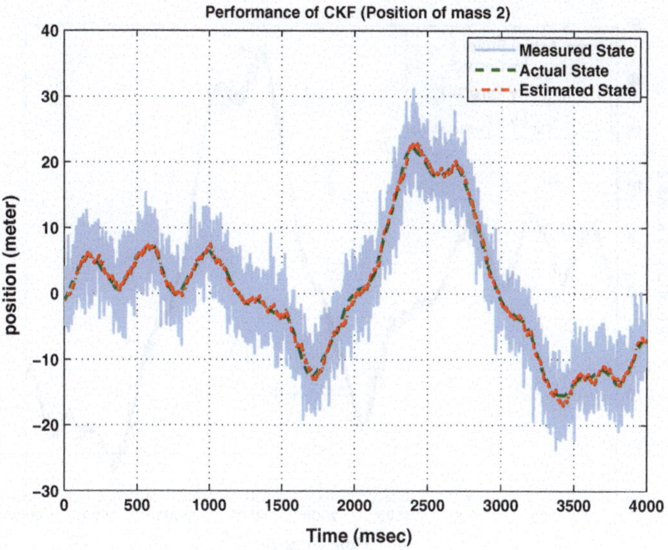

Fig. 7.6 Performance of KF without measurement loss

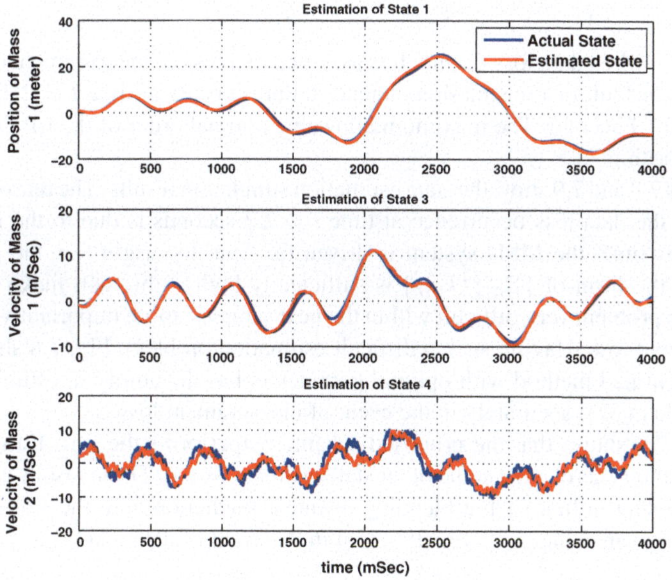

Fig. 7.7 KF without measurement loss for states x_1, x_3 and x_4

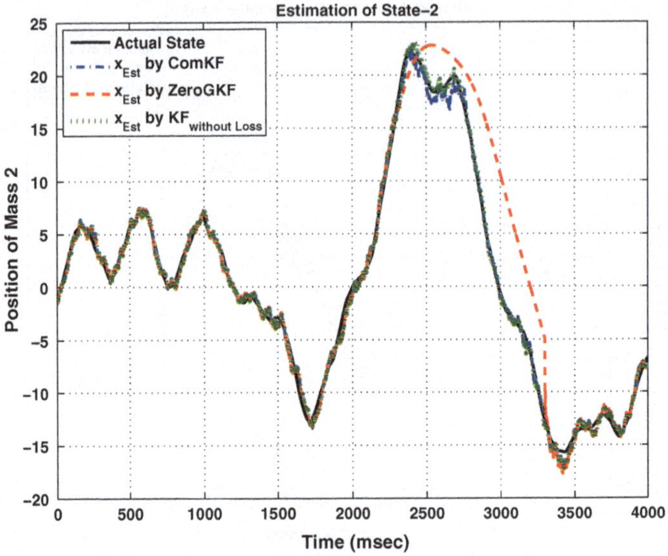

Fig. 7.8 Estimation of state x_2

(ZeroGKF), when data loss occurs. It is assumed that loss of measurement data (due to the sensor fault or transmission channel failure) occurs at time $t = 2.3$ seconds and lasts for 1 second. The maximum allowable optimal order of the LP filter M is 30 in Algorithm 7.4.

Figures 7.8 and 7.9 show the state estimation simulation results. The reason behind enforcing the data loss occurrence at time $t = 2.3$ seconds is due to the fact that, around this time, the MSD system is driven far from its origin (i.e. the potential energy in the form of $\{\Sigma_{i=1}^{i=2}\frac{1}{2}k_i x_i^2\}$ is sufficiently high)—this will make the state estimation problem very difficult within the next samples, more importantly with the measurement loss. Having such a difficult estimation problem, Fig. 7.8 shows that the compensated method with optimal parameters has the ability in estimating the output state ($x_2(t)$) accurately in the event of measurement loss.

Figure 7.9 shows that the proposed scheme outperforms the zero Kalman gain matrix method (ZeroGKF) at the other state estimations too. These results are in fact very promising in that such a blending of linear prediction with the Kalman filter would lead to an enhanced estimation tool in the event of data loss.

State Estimation Errors

Figures 7.10 and 7.11 illustrate the superiority of the ComKF over the ZeroGKF by manifesting the error analysis for each state. It can be seen that during the LOOB, the state estimation error using the ZeroGKF approach exceeds the limited boundaries excessively. On the other hand, the error generated by the proposed ComKF approach

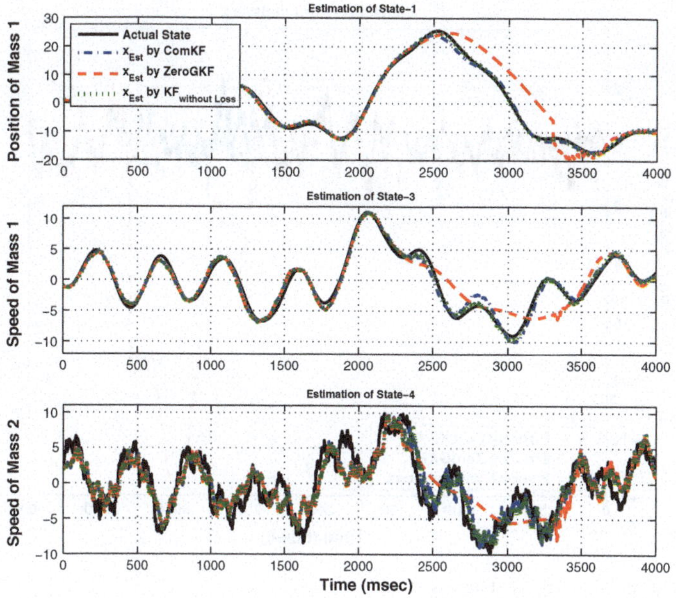

Fig. 7.9 Estimation analysis of states x_1, x_3 and x_4

is significantly smaller and hence it is less influential. This is in fact one of the major achievements one could experience when applying ComKF into their applications.

Kalman Filter Gains

The Kalman filter gains computed based on the ZeroGKF and ComKF approaches are shown in Figs. 7.12 and 7.13, respectively. As shown in Fig. 7.12, Kalman filter gains reached their steady state values in the failure-free case (no LOOB) in about 2 seconds. In the event of LOOB, the gains are forces to zero until LOOB is recovered with the zeroGKF approach. It is also worthwhile to stress that at the time of switching from ZeroGKF back to the normal Kalman filter, there are undesirable, excessive oscillations (also called the wind-up) in the gains to help hold the system's state swiftly. Kalman filter gains using ComKF are shown in Fig. 7.13 and compared with that of the normal operation. Due to the gains being calculated online, it can be seen that when the LOOB occurs, the compensated Kalman filter gains are very close to those of the original estimation system. This indicates that the proposed filter, ComKF, could successfully predict the state estimation of the original systems even if such a long period of loss of observation has occurred. Clearly, such an optimal compensated filter will generate minimum errors for the estimated vector.

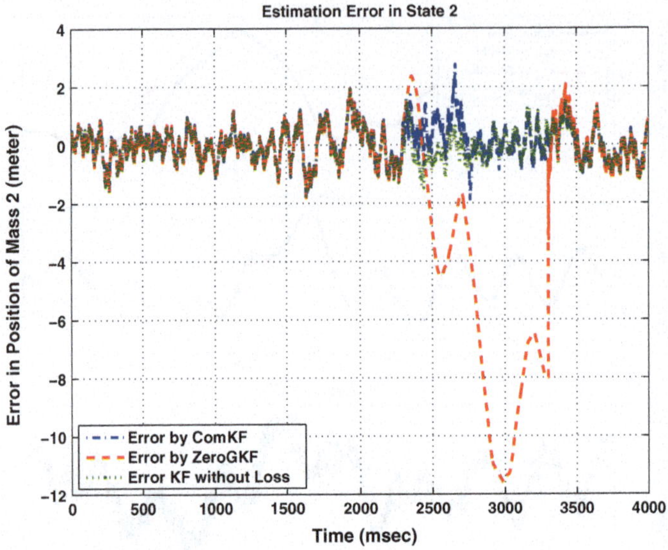

Fig. 7.10 Estimation error of state x_2

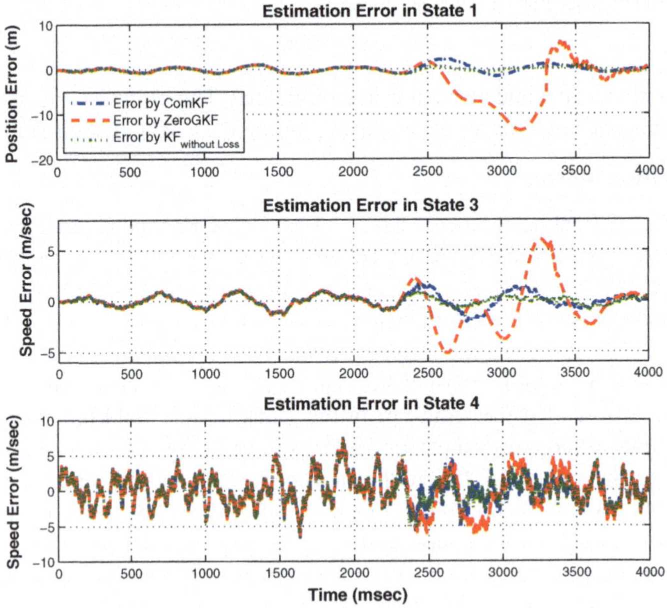

Fig. 7.11 Estimation error of states x_1, x_3 and x_4

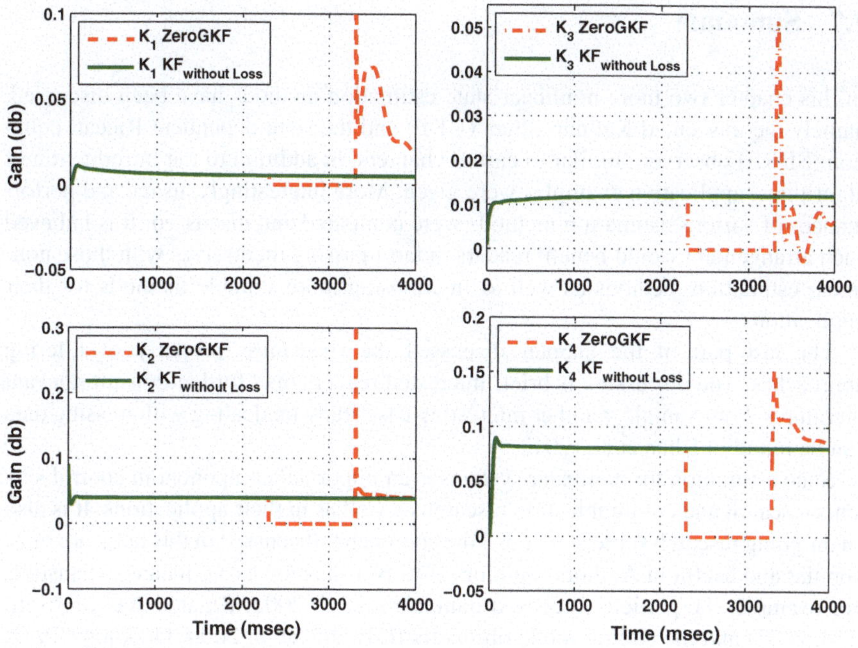

Fig. 7.12 Kalman filter gains of ZeroGKF approach

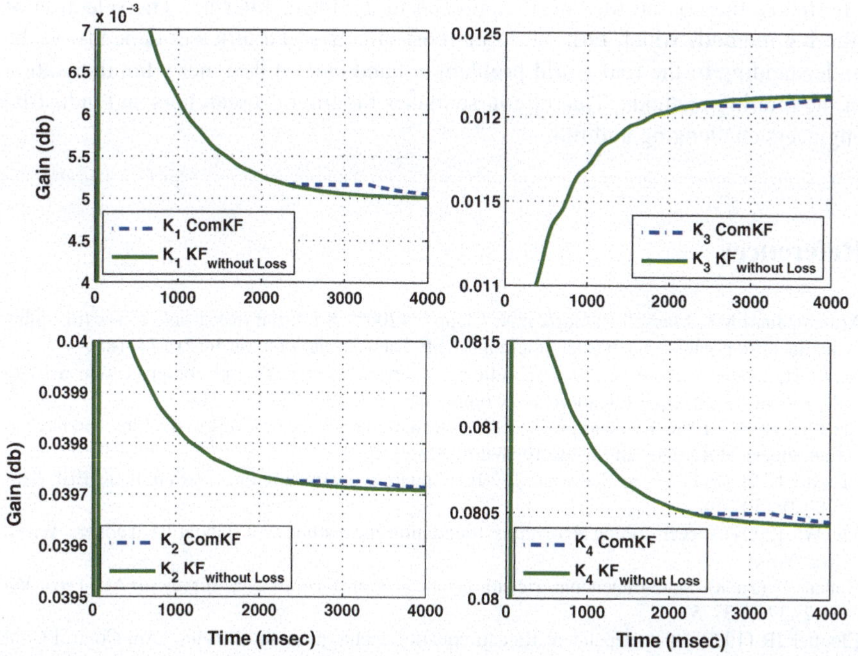

Fig. 7.13 Kalman filter gains of ComKF approach

7.7 Summary

In this chapter two more nonlinear state estimation methods have been discussed, namely the unscented Kalman filters (UKF) and the state-dependent Riccati equation (SDRE) observer. Similar to earlier chapters, in addition to the introduction of algorithms, application examples were given. More interestingly, usage and performances of various estimation methods were compared and discussed. It is believed such arrangement would benefit readers in familiarising themselves with those nonlinear estimation methods as well as in choosing more suitable methods for their applications.

The last part of the chapter discussed the robustness property of filtering approaches. The discussion is brief. Interested readers may further consult relevant literatures. For example, another interesting case study on dealing with missing data can be found in Khan et al. (2012).

State estimation for nonlinear systems is an important component in control systems research and communication research as well as in their applications. It is also an on-going research subject area. All the approaches discussed in this book are most popular and useful in Authors' opinion. However they are by no means exhaustive. For example, the particle filters (Arulampalam et al. 2002; Rigatos 2009; Grisetti et al. 2007) and the sliding-mode observers (Edwards et al. 2000; Drakunov 1992; Shtessel et al. 2014) have been proposed and researched for the state estimation problems for nonlinear control systems. Like in almost all engineering and scientific fields, there is no the "best" approach in nonlinear filtering. The selection of filtering methods which turn out to be most effective and efficient depends on the understanding of the real-world problem at hand and on designer's familiarisation of the filtering methods. That of course makes the life of researchers and industrial engineers challenging and fun.

References

Arulampalam MS, Maskell S, Gordon N, Clapp T (2002) A tutorial on particle filters for online nonlinear/non-gaussian bayesian tracking. IEEE Trans Signal Process 50(2):174–188

Banks H, Lewis B, Tran H (2007) Nonlinear feedback controllers and compensators: a state-dependent Riccati equation approach. Comput Optim Appl 37(2):177–218

Bar-Shalom, Y, Li XR, Kirubarajan T (2004) Estimation with applications to tracking and navigation: theory algorithms and software. Wiley, New York

Chandra KPB, Gu D-W, Postlethwaite I (2013) Square root cubature information filter. IEEE Sens J 13(2):750–758

Chu WC (2004) Speech coding algorithms: foundation and evolution of standardized coders. Wiley, New York

Cimen T (2008) State-dependent riccati equation (sdre) control: a survey. IFAC Proc Vol 41(2):3761–3775

Cloutier JR (1997) State-dependent Riccati equation techniques: an overview. Am Control Conf 2:932–936

Drakunov SV (1992) Sliding-mode observers based on equivalent control method. In Proceedings of the 31st IEEE conference on decision and control, 1992. IEEE, pp 2368–2369

Edwards C, Spurgeon SK, Patton RJ (2000) Sliding mode observers for fault detection and isolation. Automatica 36(4):541–553

Fekri S, Athans M, Pascoal A (2007) Robust multiple model adaptive control (RMMAC): a case study. Int J Adapt Control Signal Process 21(1):1–30

Grisetti G, Stachniss C, Burgard W (2007) Improved techniques for grid mapping with rao-blackwellized particle filters. IEEE Trans Robot 23(1):34–46

Gu D-W, Petkov PH, Konstantinov MM (2014) Robust control design with MATLAB®. Springer Science & Business Media

Hayes, MH (2009) Statistical digital signal processing and modeling. Wiley, New York

Jaganath C, Ridley A, Bernstein D (2005) A SDRE-based asymptotic observer for nonlinear discrete-time systems. Proc Am Control Conf 2005:3630–3635

Julier SJ, Uhlmann JK (2004) Unscented filtering and nonlinear estimation. Proc IEEE 92(3):401–422

Julier S, Uhlmann J, Durrant-Whyte HF (2000) A new method for the nonlinear transformation of means and covariances in filters and estimators. IEEE Trans Autom Control 45(3):477–482

Kanjilal PP (1995) Adaptive prediction and predictive control, vol 52. IET

Khan N (2012) Linear prediction approaches to compensation of missing measurements in kalman filtering. PhD thesis, University of Leicester

Khan N, Gu D-W (2009) Properties of a robust kalman filter. IFAC Proc Vol 42(19):465–470

Khan N, Fekri S, Gu D (2010) Improvement on state estimation for discrete-time LTI systems with measurement loss. Measurement 43(10):1609–1622

Khan N, Khattak MI, Gu D (2012) Robust state estimation and its application to spacecraft control. Automatica 48(12):3142–3150

Lam Q, Xin M, Cloutier J (2012) SDRE control stability criteria and convergence issues: where are we today addressing practitioners' concerns? In Infotech@ Aerospace 2012, p 2475

Liang Y-W, Lin L-G (2013) Analysis of SDC matrices for successfully implementing the SDRE scheme. Automatica 49(10):3120–3124

Liu X, Goldsmith A (2004) Kalman filtering with partial observation losses. In 43rd IEEE conference on decision and control, 2004. CDC, vol 4. IEEE, pp 4180–4186

Makhoul J (1975) Linear prediction: a tutorial review. Proc IEEE 63(4):561–580

Mracek CP, Clontier J, D'Souza CA (1996) A new technique for nonlinear estimation. In Proceedings of the 1996 IEEE international conference on control applications, 1996. IEEE, pp 338–343

Mutambara AG (1998) Decentralized estimation and control for multisensor systems. CRC Press, Boca Raton

Raol J, Girija G (2002) Sensor data fusion algorithms using square-root information filtering. IEE Proc-Radar Sonar Navig 149(2):89–96

Rigatos GG (2009) Particle filtering for state estimation in nonlinear industrial systems. IEEE Trans Instrum Meas 58(11):3885–3900

Schenato L (2008) Optimal estimation in networked control systems subject to random delay and packet drop. IEEE Trans Autom Control 53(5):1311–1317

Shamma JS, Cloutier JR (2003) Existence of sdre stabilizing feedback. IEEE Trans Autom Control 48(3):513–517

Shi L, Epstein M, Tiwari A, Murray RM (2005) Estimation with information loss: asymptotic analysis and error bounds. In 44th IEEE conference on decision and control, 2005 and 2005 European control conference. CDC-ECC'05. IEEE, pp 1215–1221

Shtessel Y, Edwards C, Fridman L, Levant A (2014) Sliding mode control and observation, vol 10. Springer, Berlin

Simon D (2006) Optimal state estimation: Kalman, H infinity, and nonlinear approaches. Wiley, New York

Sinopoli B, Schenato L, Franceschetti M, Poolla K, Jordan MI, Sastry SS (2004) Kalman filtering with intermittent observations. IEEE Trans Autom Control 49(9):1453–1464

Vaidyanathan P (2007) The theory of linear prediction. Synth Lect Signal Process 2(1):1–184
Vaidyanathan P (2008) The theory of linear prediction. Synth Lect Signal Process 2(1):1–184
Van Der Merwe R (2004) Sigma-point Kalman filters for probabilistic inference in dynamic state-space models

Index

A
Autocorrelation, 17

B
Brushless DC motor, 83

C
Closed-loop system, 4
Continuous stirred tank reactor (CSTR), 143
Continuous-time system, 3
Control plant, 2
Control process, 2
Control system, 2
Convergence, 21
Correlation, 17
Covariance, 16
Cubature $\mathcal{H}_\infty$ filter, 107
Cubature $\mathcal{H}_\infty$ information filter, 109
Cubature information filter, 100
Cubature Kalman filter, 7, 75
Cubature transformation, 80

D
DC motor, 25, 65
Discrete-time system, 3
Discrete-time system model, 30, 37
Discrete-time system model-information filters, 98

E
Extended $\mathcal{H}_\infty$ filter, 59, 64
Extended $\mathcal{H}_\infty$ information filter, 105
Extended information filter, 61, 99
Extended Kalman filter, 7, 59, 60
Extended Kalman information filter, 59

F
Feedback system, 4
Filtering, 13
Filters, 13

G
Gaussian distribution, 18

H
$\mathcal{H}_\infty$ filter, 37

I
Information filter, 6, 34, 97
Input, 2

K
Kalman filter, 5, 6, 29

L
Least square estimator, 22, 23
Linear system, 3, 22
Luenberger state observer, 24

M
Mean, 15
Mean square error, 21
MIMO system, 2
Multivariable system, 3

K. P. B. Chandra and D.-W. Gu, *Nonlinear Filtering*,
https://doi.org/10.1007/978-3-030-01797-2

N
Noise, 3
Nonlinear system, 3, 22
Normal distribution, 18

O
Open-loop system, 4
Output, 2

P
Parameter estimation, 22
Permanent magnet synchronous motor (PMSM), 129
Prediction, 13

Q
Quadruple-tank system, 39

R
Random process, 17, 19
Random variable, 14, 30
Robustness, 22

S
Sensitivity, 22
Simultaneous localisation and mapping (SLAM), 68
SISO system, 2
Smoothing, 13
Square-root cubature $\mathcal{H}_\infty$ filter, 120
Square-root cubature $\mathcal{H}_\infty$ information filter, 124
Square-root cubature information filter, 118
Square-root extended information filter, 113
Square-root extended Kalman filter, 112
Square-root filter, 7
Square-root $\mathcal{H}_\infty$ filter, 114
Stability, 21
Standard deviation, 16
State-dependent coefficient (SDC), 155
State-dependent Riccati equation (SDRE) observer, 155
State estimation, 2, 5, 13
Stationary process, 20
System model, 3, 22
System state, 1, 13

T
Time-invariant system, 3

U
Unscented Kalman filter (UKF), 7, 149, 153
Unscented transformation, 149

V
Variance, 16

W
White noise, 20

The manufacturer's authorised representative in the EU is Springer Nature Customer Service Centre GmbH, Europaplatz 3, 69115 Heidelberg, Germany. If you have any concerns regarding our products, please contact ProductSafety@springernature.com

Printed and bound by CPI Group (UK) Ltd, Croydon, CR0 4YY
05/10/2026
02228145-0002